Adaptation Measures for Urban Heat Islands

Adaptation Measures for Urban Heat Islands

Edited by

Hideki Takebayashi

Associate Professor
Department of Architecture
Kobe University, Kobe
Japan

Masakazu Moriyama

Professor emeritus
Engineering
Kobe University, Kobe
Hyogo, Japan

Academic Press is an imprint of Elsevier
125 London Wall, London EC2Y 5AS, United Kingdom
525 B Street, Suite 1650, San Diego, CA 92101, United States
50 Hampshire Street, 5th Floor, Cambridge, MA 02139, United States
The Boulevard, Langford Lane, Kidlington, Oxford OX5 1GB, United Kingdom

Library of Congress Cataloging-in-Publication Data
A catalog record for this book is available from the Library of Congress

British Library Cataloguing-in-Publication Data
A catalogue record for this book is available from the British Library

ISBN: 978-0-12-817624-5

For information on all Academic Press publications visit our website at https://www.elsevier.com/books-and-journals

Publisher: Joe Hayton
Acquisitions Editor: Graham Nisbet
Editorial Project Manager: Ali Afzal-Khan
Production Project Manager: Poulouse Joseph
Cover Designer: Alan Studholme

Typeset by TNQ Technologies

Contents

Contributors

Hiroyuki Akagawa, Ph.D
Senior Engineer, Technical Research Institute, Obayashi Corporation, Tokyo, Japan

Shinichi Kinoshita, Dr.
Associate Professor, Department of Mechanical Engineering, Osaka Prefecture University, Sakai, Japan

Noboru Masuda, Dr.
Professor Emeritus, Director of Plant Factory Research and Development Center, Osaka Prefecture University, Sakai, Japan

Keiko Masumoto, Dr.
Researcher, Research Group, Osaka City Research Center of Environmental Science, Osaka, Japan

Ikusei Misaka, Dr.
Professor, Section of Architecture, Nippon Institute of Technology, Saitama, Japan

Masakazu Moriyama, Dr.
Professor Emeritus, Department of Architecture, Kobe University, Kobe, Japan

Minako Nabeshima, Ph.D
Professor, Department of Urban Design and Engineering, Osaka City University, Osaka, Japan

Nobuya Nishimura, Dr.
Professor, Department of Mechanical and Physical Engineering, Osaka City University, Osaka, Japan

Toru Shiba
Manager, Residential Energy Business Unit, OSAKAGAS CO.,LTD., Osaka, Japan

Hideki Takebayashi, Dr.
Associate Professor, Department of Architecture, Kobe University, Kobe, Japan

Shinji Yoshida, Ph.D
Associate Professor, Department of Residential Architecture and Environmental Science, Nara Women's University, Nara, Japan

Atsumasa Yoshida, Dr.
Professor, Department of Mechanical Engineering, Osaka Prefecture University, Sakai, Japan

CHAPTER

Background and purpose

1

Hideki Takebayashi, Dr. [1], **Masakazu Moriyama, Dr.** [2]

[1] *Associate Professor, Department of Architecture, Kobe University, Kobe, Japan;* [2] *Professor Emeritus, Department of Architecture, Kobe University, Kobe, Japan*

Chapter outline

1. Background

Mitigation measures, such as green roof, cool roof (with a high reflectance material), and water-retentive materials, have been developed with an expectation that they will serve as countermeasures to the urban heat island [1–4]. In recent years, adaptation measures, such as awnings, louvers, directional reflective materials, mist, and evaporative materials, have been developed with the expectation that they will serve as effective solutions to outdoor human thermal environments that are under the influence of urban heat islands.

The Japanese Ministry of the Environment developed the "Heat countermeasure guideline in the city" [5], which includes basic, specific adaptation measures, and technical sections. The guideline states that "by understanding the factors that make it hot and implementing appropriate adaptation measures for places we have to wait for or places we want to spend comfortably such as bus stops and plazas, we can promote a healthy and comfortable environment in the urban area." In the basic section, the adaptation measures against heat are explained in an accessible manner for the Japanese administration and the general public. In the specific adaptation measures section, the type and effect of adaptation measure technologies and precautions to be considered upon introducing them are explained for the general public and practitioners involved in town development. In the technical section,

Adaptation Measures for Urban Heat Islands. https://doi.org/10.1016/B978-0-12-817624-5.00001-4

technical information on adaptation measure technologies is explained for building and external construction design practitioners.

Several studies focused on effective measures against heat waves have been implemented in various countries [6,7]. Evaporative cooling effects such as irrigation [6,7], vegetation, and pavement watering [7] have been studied by the numerical simulation. Some of those scenarios such as evaporative cooling and greening are assumed to future climate affected by climate change [7,8]. Discussions including the improvement of thermal environment in the street canyon or in the plaza were not sufficiently conducted in previous examinations [9–11].

2. Adaptive city strategy in Karlsruhe [12]

In Germany, several cities are considering adaptation measures. According to a report from Karlsruhe City [12], it is recommended that appropriate adaptation measures be introduced in "hot spots" where temperatures are high. Several typical urban districts in cities that may undergo adaptation in the future are also discussed. The framework plan on the adaptive city strategy in Karlsruhe has been decided for the hot thermal environment. Fig. 1.1 shows the various countermeasure menus for the climate change adaptive city that has been filled under the space scale. Fig. 1.2 shows an example of the city block. Fig. 1.2A shows the current situation of the city block with land cover. Fig. 1.2B shows the state of future situation in 2050 after the countermeasures to the hot thermal environment.

3. Overview of efforts in Osaka

3.1 The efforts by Osaka local government

Osaka heat island measures promotion plan was established in 2004 by Osaka Prefecture [13]. At the beginning of the plan, the current situation, causes, and effects of urban heat islands were described based on statistical data. In Osaka, air temperature rises by 2.1 degrees in 100 years, which exceeds the national average of 1.0 degrees, and the difference of 1.1 degrees is considered to be the effect by urban heat islands. Looking at the change in the land use condition in Osaka area, forests and wild areas decreased by 11% and farmland by 42%, residential areas increased by 34% and roads by 50%, in the last 30 years. Simultaneously, the total energy consumption in Osaka area increased by 25%. In Osaka city, the number of carriers due to heatstroke increased in the year the number of higher air temperature hours was large. The basic concepts of the plan are organized as follows.

- A town where we can comfortably live and walk
- A town full of green, water, and natural winds
- A town where we can feel coolness and freshness
- A town where we can sleep without relying on air conditioner
- A town which does not waste energy
- A town where so much heat is not accumulated on buildings and roads

Measures

HIGHER LEVEL MEASURES

M 01
Preservation and creation of cold air production areas and cold air paths

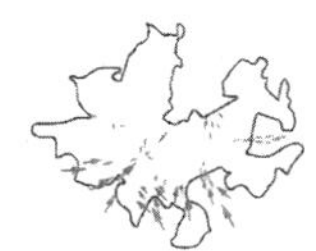

M 02
Preservation and creation of large green areas (open space and forest)

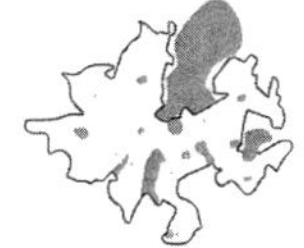

M 03
Connecting of open space (city parks) and larger green spaces

M 04
Reduction of anthropogenic heat emissions

M 05
Preservation and creation of open and flowing water areas

LOCAL LEVEL MEASURES

M 06
Demolition (decongestion)

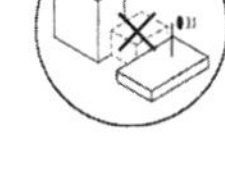

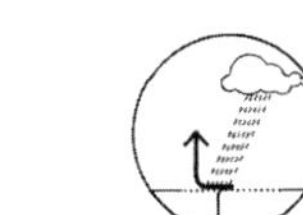

M 07
Unsealing

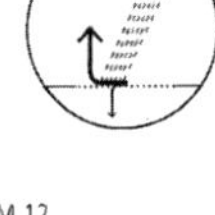

M 08
Green parking and shading parking

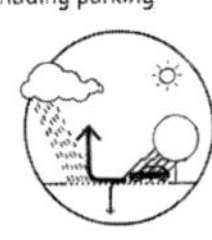

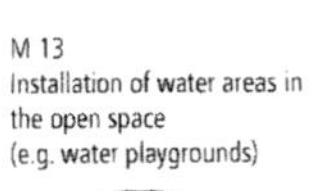

M 09
Shading of streets, places and buildings

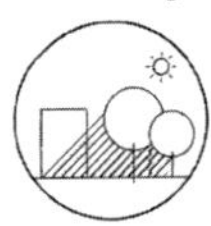

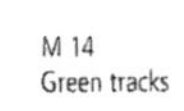

M 10
New installation of pocket-parks

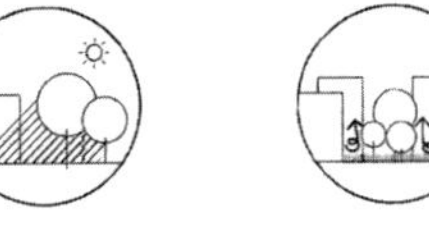

M 11
Greening of courtyards and backyards

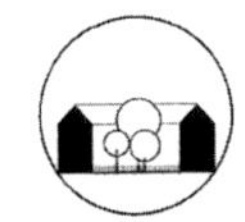

M 12
Increase of surface albedo (reflection)

M 13
Installation of water areas in the open space (e.g. water playgrounds)

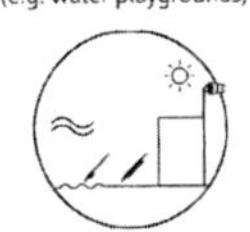

M 14
Green tracks

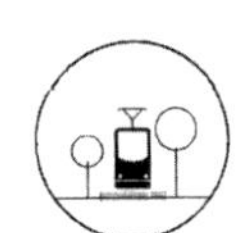

MEASURES ON BUILDINGS

M 15
Energy-related modernisation of building

M 16
Roof greening

M 17
Façade greening

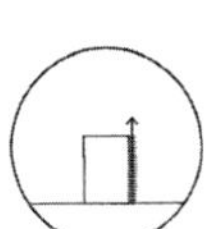

M 18
Summer thermal insulation of buildings

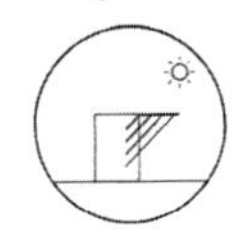

SUPPLEMENTARY SOCIAL MEASURES

M 19
Neighborly help projects

FIGURE 1.1

Various countermeasure menus for the climate change adaptive city ([12] appended map).

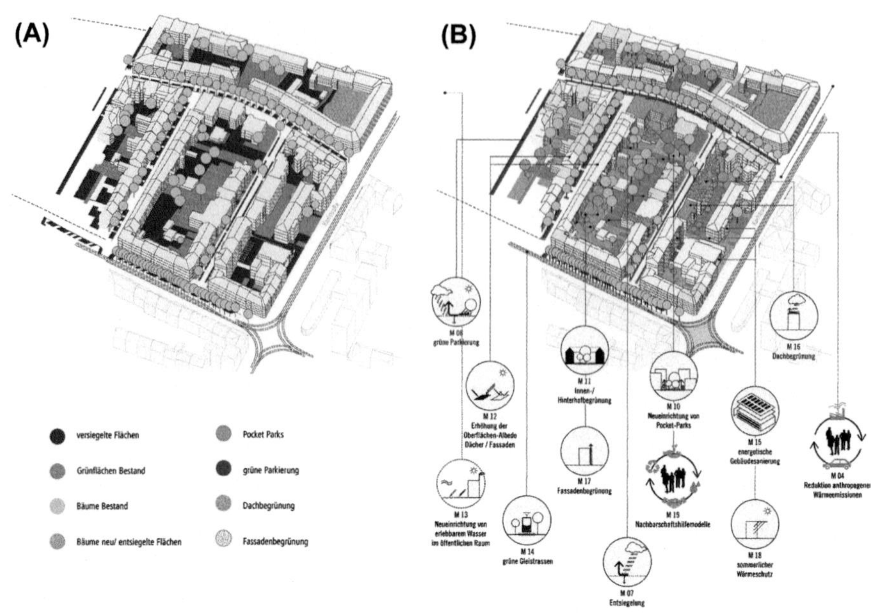

FIGURE 1.2

Example of the city block. (A) Current situation, (B) Future situation in 2050 after the countermeasures ([12] p. 52 and p. 53).

Plan period is until 2025; the objective area of the plan is the whole of Osaka Prefecture. The goals of the plan are the following two.

Goal 1 Reduce nighttime air temperature in residential areas and reduce the number of tropical nights by 30% by 2025. Tropical night is a night when minimum air temperature is 25°C or more.

Goal 2 Create cool spots in the outdoor space, improve the thermal environment during the summer daytime, and lower the sensational temperature of the human body.

The basic directions of measures are the following three.

1) Reduction of anthropogenic heat release by introduction of energy saving equipment and improvement of lifestyle

2) Control of surface temperature rise by improvement of surface coverage of lands and buildings

3) Utilization of the cooling effect of wind, green and water

Osaka heat island measures promotion plan was revised in 2015 by Osaka Prefecture and Osaka City [14]. According to statistical data, the number of tropical nights decreased slightly in recent years, but the number of heatstroke patients due to higher air temperatures during the day exceeded 4000 in 2013, and daytime health damage became significant in recent years. Therefore, in addition to conventional summer nighttime measures, mitigation of daytime heat stress, that is, promotion of adaptation measures, was added to the plan. The basic concepts of the plan are organized as follows.

- Steady promotion of "mitigation measures" which include efforts to reduce the surface temperature of building and ground surfaces and reduce anthropogenic heat release
- In addition to "mitigation measures" promoting "adaptation measures" an effort to reduce the impact on human health, etc.
- Especially in the downtown area of Osaka, implementing various measures for redevelopment of urban areas and urban infrastructure
- In order to reduce the number of tropical nights, properly progress management by using appropriate indicators

Plan period is the same as before, until 2025. The goals of the plan are also the same as before, the following two.

Goal 1 Reduce nighttime air temperature in residential areas and reduce the number of tropical nights by 30% by 2025.

Goal 2 Create cool spots in the outdoor space, improve the thermal environment during the summer daytime, and lower the sensational temperature of the human body.

A fourth one was added to the previous three for the basic directions of measures.

1) Reduction of anthropogenic heat release
2) Control of surface temperature rise
3) Utilization of the cooling effect of wind, green, and water
4) Promotion of adaptation measures

3.2 The efforts by Osaka HITEC

Osaka Heat Island Countermeasure Technology Consortium (HITEC) was established in January 2006, for the purpose of the development and spread of heat island countermeasure technologies, implementation of measures and verification of their effects, and the collaboration between industry, academia, government, and the private sectors [15]. The consortium called on local governments, private companies, universities, research institutes, environmental NGOs, and NPOs to participate. The consortium has five working groups (WGs); WG related to the materials, WG on heat utilization and reduction of anthropogenic heat release, WG on cool spot creation technologies, WG on thermal load evaluation method, and WG on urban design.

Osaka HITEC started the certification system of heat island measures technology in October 2011. In 2019, the category of certification technology increased to nine; high solar reflectance paint for roof, high solar reflectance pavement (excluding for roadways), high solar reflectance waterproof sheet (membrane), high solar reflectance roofing materials (tile, slate, metal, etc.), water-retaining pavement block, external insulation specification for roof, external insulation specification for outer wall, retroreflective high solar reflectance outer wall material, and retroreflective high solar reflectance window film. Three high solar reflectance paints for roof, five high solar reflectance pavements (excluding for roadways), three high solar reflectance roofing materials, one external insulation specification for outer wall, and one retroreflective high solar reflectance window film were certified [15].

Osaka HITEC has also held town planning ideas competition in consideration of the heat islands measures, recruitment and selection of cool spots and cool roads, and technology seminars.

Osaka HITEC is currently working on evaluation and implementation of adaptation measures against extreme heat. Several adaptation technologies have been developed by various companies and their evaluation methods were discussed so that they may be properly implemented in society. Adaptation measures for urban heat islands and their effects and associated evaluation indices are organized. A simple method to evaluate the adaptation measures is examined focusing on their appropriate introduction in urban space. Osaka HITEC is promoting the following three activities; organization of adaptation measures, examination of specific image of adaptive city, and examination of evaluation method of adaptive city. In this book, the results of above activities are organized. The table of contents and the outline of this book are as follows.

1. Background and purpose (by Prof. Takebayashi and Prof. Moriyama): Various approaches related to adaptation measures, for example, adaptation city in Karlsruhe and countermeasure guideline for heat in the urban area by Ministry of Environment of Japan are reviewed.
2. Adaptation measures and their performance (by Prof. Takebayashi, Prof. Misaka and Dr. Akagawa): Adaptation measure technologies listed in the guideline of Ministry of Environment are reviewed. These technologies are evaluated using indicators such as solar transmittance and solar reflectance.
3. Priority introduction place "hot spot" of adaptation measures (by Prof. Takebayashi): The relationship between urban morphology and solar shielding and urban ventilation is described. The appearance of hot spots in various urban blocks is presented.
4. Case study of adaptation city (by Prof. Masuda, Prof. Nishimura, Prof. Nabeshima, Mr. Shiba and Dr. Akagawa): Case studies of adaptation cities are introduced for various urban blocks.
5. Evaluation method of adaptation city (by Prof. S. Yoshida, Prof. A. Yoshida and Prof. Kinoshita): The method of human thermal environment assessment is explained. A method for simulating the thermal environment of an adaptive city is explained.
6. The role of local government (by Dr. Masumoto): In discussing adaptation cities, the role of local government is described.
7. Summary (by Prof. Takebayashi and Prof. Moriyama): Strategies for introducing adaptation measures are stated.

4. Summary

Several studies focused on effective measures against heat waves have been implemented in various countries. For example, in Germany, several cities are considering adaptation measures. According to a report from Karlsruhe City, it is recommended that appropriate adaptation measures be introduced in "hot spots" where temperatures are high. Several typical urban districts in cities that may undergo adaptation in the future are also discussed. In this book, the three activities authors have been working on in Osaka HITEC are organized; organization of adaptation measures, examination of specific image of adaptive city, and examination of evaluation method of adaptive city.

References

[1] Akbari H, Kolokotsa D. Three decades of urban heat islands and mitigation technologies research. Energy and Buildings 2016;133:834–42.

[2] Aleksandrowicz O, Vuckovic M, Kiesel K, Mahdavi A. Current trends in urban heat island mitigation research: observations based on a comprehensive research repository. Urban Climate 2017;21:1–26.

[3] Santamouris M. Cooling the cities – a review of reflective and green roof mitigation technologies to fight heat island and improve comfort in urban environments. Solar Energy 2014;103:682–703.

[4] Santamouris M. Using cool pavements as a mitigation strategy to fight urban heat island—a review of the actual developments. Renewable and Sustainable Energy Reviews 2013;26:224–40.

[5] The Ministry of the Environment of Japan. Heat countermeasure guideline in the city. 2018. http://www.env.go.jp/air/life/heat_island/guidelineH30/gudelineH30_all.pdf. [Accessed 18 December 2011].

[6] Broadbent AM, Coutts AM, Tapper NJ, Demuzere M. The cooling effect of irrigation on urban microclimate during heatwave conditions. Urban Climate 2018;23:309–29.

[7] Daniel M, Lemonsu A, Viguié V. Role of watering practices in large-scale urban planning strategies to face the heat-wave risk in future climate. Urban Climate 2018;23: 287–308.

[8] de Munck C, Lemonsu A, Masson V, Le Bras J, Bonhomme M. Evaluating the impacts of greening scenarios on thermal comfort and energy and water consumptions for adapting Paris city to climate change. Urban Climate 2018;23:260–86.

[9] Baklanov A, Grimmond CSB, Carlson D, Terblanche D, Tang X, Bouchet V, Lee B, Langendijk G, Kolli RK, Hovsepyan A. From urban meteorology, climate and environment research to integrated city services. Urban Climate 2018;23:330–41.

[10] Gao Z, Bresson R, Qu Y, Milliez M, Munck C, Carissimo B. High resolution unsteady RANS simulation of wind, thermal effects and pollution dispersion for studying urban renewal scenarios in a neighborhood of Toulouse. Urban Climate 2018;23:114–30.

[11] Ng E, Ren C. China's adaptation to climate & urban climatic changes: a critical review. Urban Climate 2018;23:352–72.

[12] Berchtoldkrass space & options - Raumplaner, Stadtplaner Partnerschaft (Martin Berchtold, Philipp Krass, Poliksen Oorri Dragaj, Maren van der Meer, Tobias Rahn, Lisa Brandstetter), Geo-Net Umweltconsulting GmbH (Peter Trute, Bjoern Bueter), Juergen Baumueller, Guenter Gross, "Staedtebaulicher Rahmenplan Klimaanpassung - Anpassungskomplex Hitze", Stadt Karlsruhe, Stadtplanungsamt (Anke Karmann-Woessner, Heike Dederer, Martin Kratz), Germany, 2015.

[13] Osaka Prefecture. Osaka heat island measures promotion plan. 2004. http://www.pref.osaka.lg.jp/chikyukankyo/jigyotoppage/heat_mati.html. [Accessed 18 June 2006].

[14] Osaka Prefecture. Revised Osaka heat island measures promotion plan. 2015. http://www.pref.osaka.lg.jp/chikyukankyo/jigyotoppage/osakaheatkeikaku.html. [Accessed 18 June 2006].

[15] Osaka Heat Island Countermeasure Technology Consortium. HITEC news. 2018. http://osakahitec.com/active/news/news2018_01_vol14.pdf. [Accessed 18 August 2009].

CHAPTER

Adaptation measures and their performance

2

Hideki Takebayashi, Dr. [1], **Ikusei Misaka, Dr.** [2], **Hiroyuki Akagawa, Ph.D** [3]

[1]*Associate Professor, Department of Architecture, Kobe University, Kobe, Japan;* [2]*Professor, Section of Architecture, Nippon Institute of Technology, Saitama, Japan;* [3]*Senior Engineer, Technical Research Institute, Obayashi Corporation, Tokyo, Japan*

Chapter outline

1. Adaptation measures

In recent years, in order to serve as effective solutions to outdoor human thermal environments under the influence of urban heat islands, adaptation measures such as awnings, louvers, directional reflective materials, mist sprays, and evaporative materials have been developed. The Japanese Ministry of the Environment

Adaptation Measures for Urban Heat Islands. https://doi.org/10.1016/B978-0-12-817624-5.00002-6

developed the "Heat countermeasure guideline in the city" [1], which includes basic, specific adaptation measures, and technical sections. The adaptation measures for urban heat island listed in the heat countermeasure guidelines established by the Japanese Ministry of Environment [1], the report by the Japanese Ministry of Environment [2], and the town planning idea competition considering the urban heat island presented at the Osaka Heat Island Countermeasure Technology Consortium [3] are shown in Table 2.1. The mechanisms by which these methods work and the evaluation index governing their effects are also presented. Heat is mainly mitigated by

Table 2.1 Adaptation measures for urban heat islands and their effects and associated evaluation indices.

Menu	Evaluation index	Main effect mechanism
From the heat countermeasure guidelines by the Japanese ministry of environment [1]		
Green shade [4]	Solar transmittance, evaporative efficiency	Sunshade, evaporative cooling
Solar radiation shade [5]	Solar transmittance, convection heat transfer coefficient	Sunshade, convection heat transfer
Retroreflective surface [6,7]	Downward solar reflectance	Solar reflection
Water-retentive pavement [8,9]	Evaporative efficiency	Evaporative cooling
Cool pavement [9]	Solar reflectance	Solar reflection
Green pavement [10]	Evaporative efficiency	Evaporative cooling
Green wall [11]	Evaporative efficiency	Evaporative cooling
Water-retentive wall [12]	Evaporative efficiency	Evaporative cooling
Fine mist spray [13,14]	Evaporation rate	Evaporative cooling
From the report by the Japanese ministry of environment [2]		
Awning [15]	Solar transmittance	Sunshade
Fractal-shaped sunshade [5]	Solar transmittance, convection heat transfer coefficient	Sunshade, convection heat transfer
Mesh shade and water supply	Solar transmittance, evaporative efficiency	Sunshade, evaporative cooling
Evaporative cooling louver [12]	Evaporative efficiency	Evaporative cooling
Greening cooling louver	Evaporative efficiency	Evaporative cooling
Tree pot	Solar transmittance, evaporative efficiency	Sunshade, evaporative cooling
Water-retentive block [8]	Evaporative efficiency	Evaporative cooling
Water surface [16]	Evaporative efficiency	Evaporative cooling
Fine mist spray with blower [13,14]	Evaporation rate	Evaporative cooling
Ceiling cooling system	Surface temperature	Artificial cooling
Water cooling bench	Surface temperature	Artificial cooling

Table 2.1 Adaptation measures for urban heat islands and their effects and associated evaluation indices.—*cont'd*

Menu	Evaluation index	Main effect mechanism
From town planning idea competition by Osaka heat island countermeasure technology consortium [3]		
Water surface [16]	Evaporative efficiency	Evaporative cooling
Watering [16]	Evaporative efficiency	Evaporative cooling
Fine mist spray [13,14]	Evaporation rate	Evaporative cooling
Shading [15]	Solar transmittance	Sunshade
Tree planting	Solar transmittance, evaporative efficiency	Sunshade, evaporative cooling
Roof and ground greening [10]	Evaporative efficiency	Evaporative cooling
Wind use	Convection heat transfer coefficient	Convection heat transfer
Traffic mode control	Anthropogenic heat release	Reduction of anthropogenic heat release
Unused energy use, natural energy use	Anthropogenic heat release	Reduction of anthropogenic heat release
ICT (Information and Communication Technology) use	Human body physiological amount	Reduction of human thermal load

solar shading, solar reflection, and evaporation. Therefore, solar transmittance, solar reflectance, and evaporative efficiency (evaporative rate) are the primary evaluation indices. The increase in the convection heat transfer coefficient is the cause of cooling by fractal-shaped sunshades, and artificial cooling is the cause of cooling by ceiling cooling systems and water cooling benches. Examples of adaptation measures developed by Japanese companies are shown in Figs. 2.1–2.3. Experiments demonstrating these measures are currently proceeding throughout Japan [1–16].

FIGURE 2.1

Automatically opening and closing awning installed at a bus stop. (A) whole view; (B) internal view; (C) closed state.

Report entrusted by the Ministry of the Environment in 2016 fiscal year; 2017 (http://www.env.go.jp/air/report/h28-02/mat09_2802.pdf).

FIGURE 2.2

Fractal-shaped sunshade. (A) in a park; (B) at a tram stop.

Report entrusted by the Ministry of the Environment in 2016 fiscal year; 2017 (http://www.env.go.jp/air/report/h28-02/mat09_2802.pdf).

FIGURE 2.3

Evaporative cooling louver. (A) in a park; (B) at a tram stop.

Report entrusted by the Ministry of the Environment in 2016 fiscal year; 2017 (http://www.env.go.jp/air/report/h28-02/mat09_2802.pdf).

2. Simple evaluation method of adaptation measures

2.1 Thermal comfort index for the evaluation of adaptation measures

The effect of the studied adaptation measures is evaluated by outdoor human thermal comfort, which is strongly correlated to the outdoor thermal environment. As Nouri et al. [17] pointed out, the selection of the index for the assessment of outdoor thermal comfort conditions is still a debated matter [18]. They also demonstrated the necessity of standardizing a thermal comfort index for specific regions. Currently, different indices are widely used by each academic community. The physiologically equivalent temperature (PET) is widely used in Europe that has been defined as the air temperature at which, in a typical indoor setting, the human energy budget is maintained by the skin temperature, core temperature, and perspiration rate, which are equivalent to those under the conditions to be assessed [19,20]. In Japan,

Table 2.2 Relationship between SET* and thermal comfort.

SET*	Thermal comfort
33.3–	Extremely uncomfortable
32.1–33.3	Uncomfortable
30.8–32.1	Slightly uncomfortable
28.4–30.8	Neither
27.0–28.4	Slightly comfortable
–27.0	Comfortable

SET*, *new standard effective temperature.*

new standard effective temperature (SET*) and wet-bulb globe temperature (WBGT) are mainly used. WBGT, which is a stress index worldwide accepted as a preliminary tool for the assessment of hot thermal environments [21–23], is often used under more severe conditions to warn of the risk of heat stroke, and SET*, defined as the equivalent dry-bulb temperature of an isothermal environment at 50% relative humidity in which a subject, while wearing clothing standardized for activity concerned, would have the same heat stress and thermoregulatory strain as in the actual test environment [24], is used to evaluate the thermal environment [1]. The relationship between SET* and thermal comfort, which is based on the results of a declaration test for the outdoor comfort of Japanese people, is shown in Table 2.2 [25]. SET* is desirable as an index from the viewpoint of appropriately introducing adaptation measures in urban areas and developing a more comfortable outdoor space as it exhibits a good relationship with outdoor thermal comfort [26].

2.2 Sensitivity analysis

Assuming a typical summer day as a standard condition; under which the air temperature is 34°C, relative humidity is 50%, wind speed is 1 m/s, mean radiant temperature (MRT) in a sunny place is 50 or 37°C in a shaded place, clothing insulation is 0.6 clo, and metabolic rate is 2 Met; SET* and PET sensitivity analysis were conducted with a variation range of 20–40°C for air temperature, 30%–80% for relative humidity, 0.5–3 m/s for wind speed, and 20–60°C for MRT [27,28].

Sensitivity analysis results are shown in Figs. 2.4 and 2.5. The sensitivities by air temperature, relative humidity, wind speed, and MRT for SET* were 0.63°C/°C, 0.13°C/%, 1.4°C/(m/s), and 0.21°C/°C, respectively. Those for PET were 0.86°C/°C, 0.01°C/%, 1.0°C/(m/s), and 0.50°C/°C, respectively. The sensitivities by MRT and wind speed for SET* were larger than those by air temperature and relative humidity; however, they were within the expected variation range of each element. For PET, the sensitivity of MRT is much higher and the sensitivity of relative humidity is much lower. PET is calculated by RayMan 1.2 by Prof. A. Matzarakis.

The relationship between air temperature, MRT, and SET*, PET are shown in Figs. 2.6 and 2.7. SET* and PET are indicated by a contour line. Above-standard conditions were set for the other elements. If the evaluation point moved from a sunny to a shaded place, the MRT decreased by 13°C and SET*, PET decreased by 2.8, 6.1°C, respectively. To obtain the same decrease in SET*, PET due to air

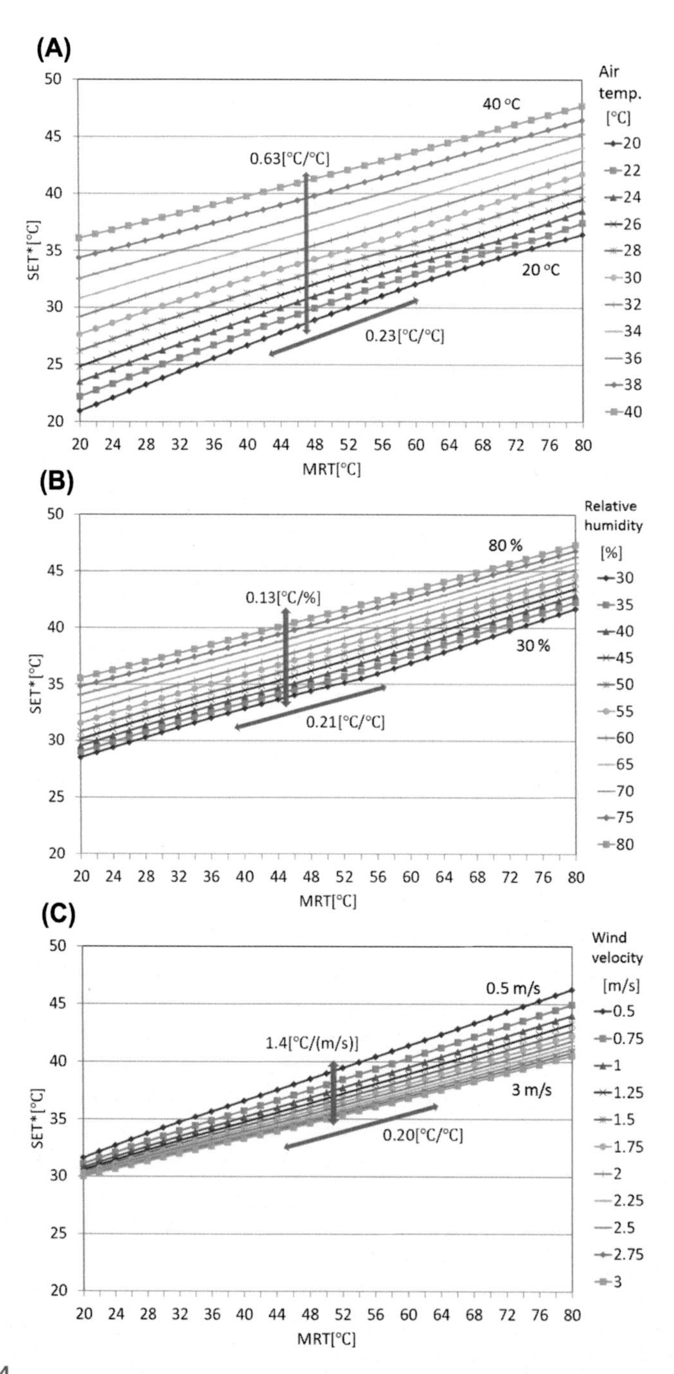

FIGURE 2.4

Sensitivity analysis results of new standard effective temperature (SET*) conducted with a variation range of (A) 20–40°C for air temperature, (B) 30%–80% for relative humidity, (C) 0.5–3 m/s for wind speed, and 20–60°C for mean radiant temperature (MRT), when clothing insulation is 0.6 clo and metabolic rate is 2 Met.

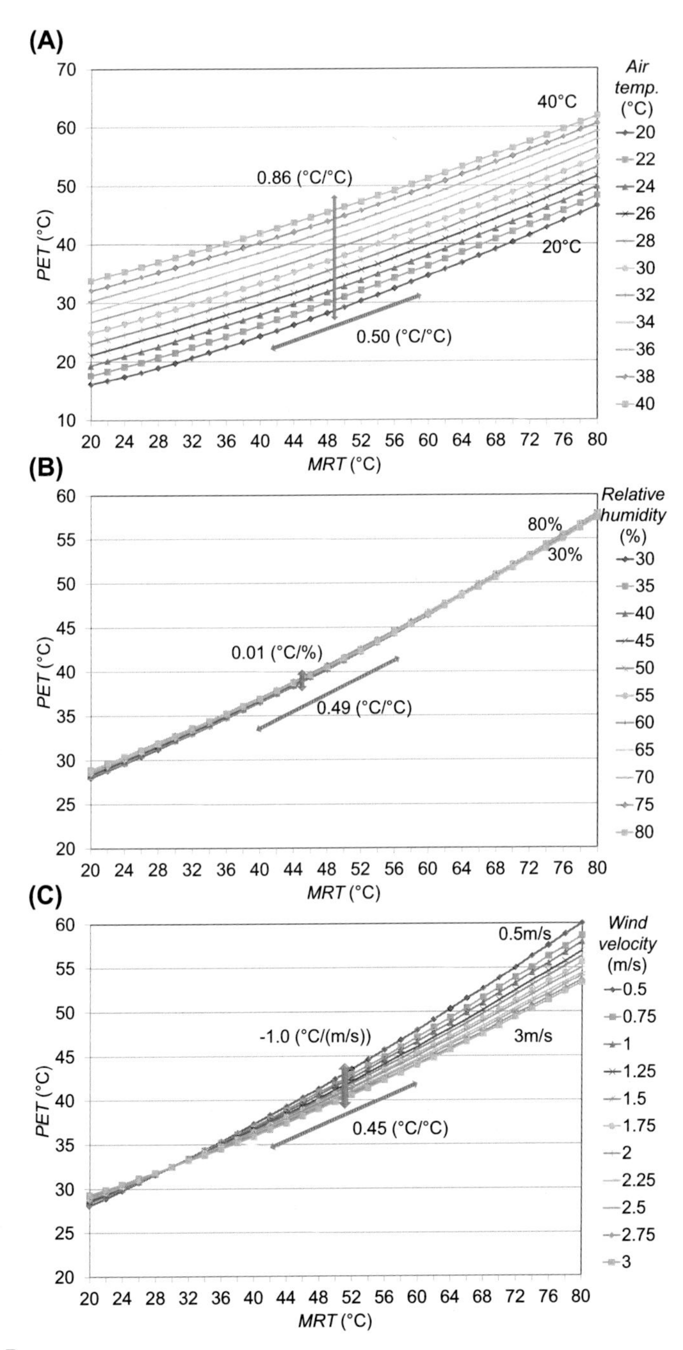

FIGURE 2.5

Sensitivity analysis results of physiologically equivalent temperature (PET) conducted with a variation range of (A) 20–40°C for air temperature, (B) 30%–80% for relative humidity, (C) 0.5–3 m/s for wind speed, and 20–60°C for MRT, when clothing insulation is 0.6 clo and metabolic rate is 2 Met.

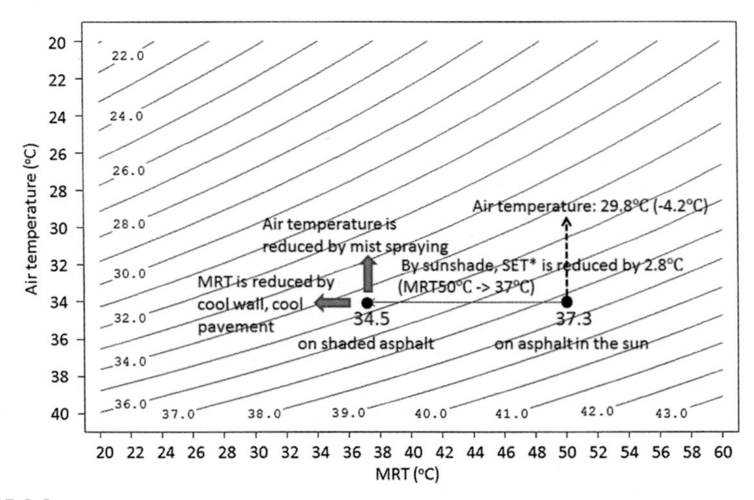

FIGURE 2.6

Relationships between air temperature, mean radiant temperature (MRT), and new standard effective temperature (SET*). SET* is indicated by a contour line. The relative humidity is 50%, wind speed is 1 m/s, clothing insulation is 0.6 clo, and metabolic rate is 2 Met.

temperature reduction by mist spraying, it must be lowered by 4.2, 7.0°C, respectively. Similarly, it is difficult to considerably reduce MRT using cool walls and pavements.

Examples of the effects of adaptation measures obtained by demonstrative experiments are shown in Figs. 2.8 and 2.9. As MRT was measured by a globe

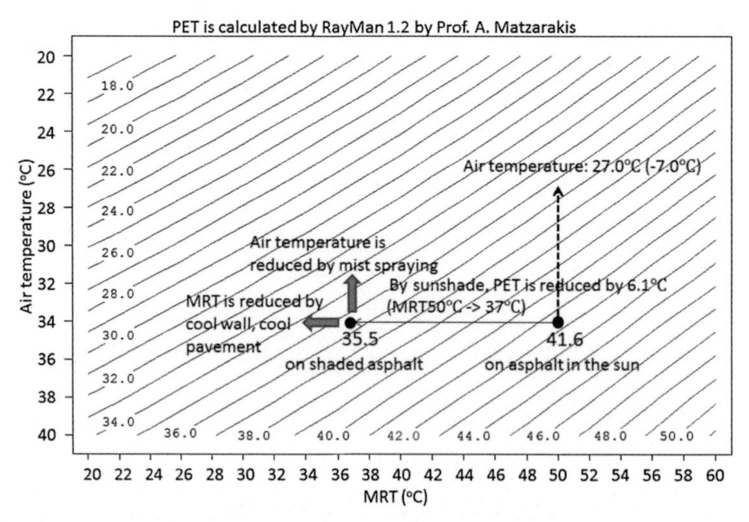

FIGURE 2.7

Relationships between air temperature, mean radiant temperature (MRT), and physiologically equivalent temperature (PET). PET is indicated by a contour line. The relative humidity is 50%, wind speed is 1 m/s, clothing insulation is 0.6 clo, and metabolic rate is 2 Met.

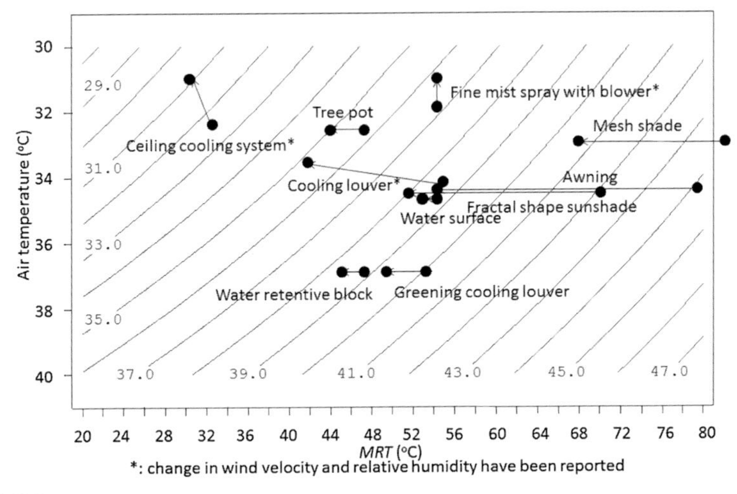

FIGURE 2.8

Examples of the effects of adaptation measures obtained through demonstrative experiments. The background new standard effective temperature (SET*) is the same as that in Fig. 2.6.

thermometer, the solar absorptance was set to 1.0, which was much larger than that of the human body. The measurements were taken at various places and times under typical summer weather condition; therefore, a simple mutual comparison was not appropriate. It was, however, possible to qualitatively recognize the characteristics of each adaptation measure [16]. Shielding of solar radiation to pedestrians was a more effective method of lowering MRT and SET*, PET.

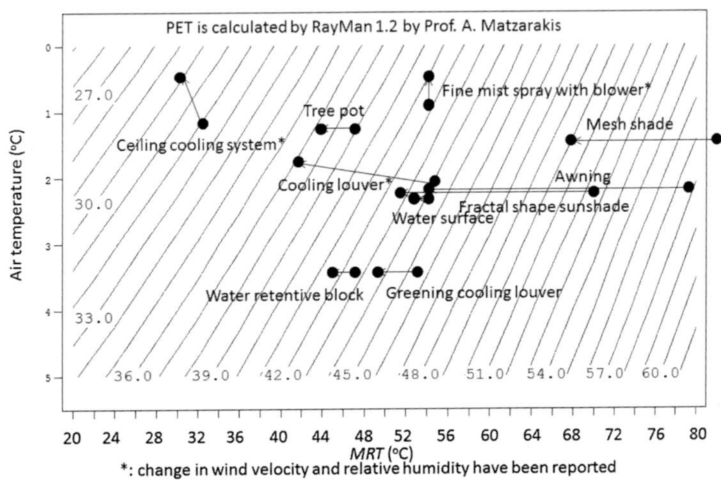

FIGURE 2.9

Examples of the effects of adaptation measures obtained through demonstrative experiments. The background physiologically equivalent temperature (PET) is the same as that in Fig. 2.7.

2.3 MRT and surface temperature reduction evaluation

The decrease in MRT caused by solar radiation shielding was dominant over the improvement in SET*, PET. Assuming the implementation of adaptation measures such as shading, MRT was evaluated using the following indices: solar transmittance τ, evaporation rate E, solar absorptance a, and convective heat transfer coefficient h. Assuming that the human body is spherical with a solar absorptance a_h which is assumed to be 0.5, MRT can be calculated from Eq. (2.1) [28–30].

$$\mathbf{MRT} = \left(a_h Q/_\sigma + \sum_{i=1} \Phi_i T_i^4 \right)^{\frac{1}{4}} \tag{2.1}$$

With reference to previous studies in Japan [27,28], the weather conditions during of typical summer day were assumed as follows; solar radiation J was 1000 W/m^2 (direct solar radiation was 900 W/m^2 and diffuse solar radiation was 100 W/m^2), each surface temperature T_i was the same as the air temperature T_a ($T_i = T_a = 34°C$), and the MRT under clear sky conditions was 56.2°C. While the relationship between the human body and the surrounding objects is varied actually, in order to simplify the discussion, it is supposed to be a human body on a green area that has been thoroughly irrigated. The incident solar radiation on the human body was calculated by $Q = 900/4 + 100$ W/m^2, as the human body was assumed to be a sphere. σ is the Stefan–Boltzmann constant ($= 5.67 \times 10^{-8}$ $W/[m^2K^4]$), and Φ_i is the shape factor between the human body and each surface.

Surface temperature T_s of the adaptation measures is calculated from Eq. (2.2).

$$T_s = \frac{1}{h}(aJ + \varepsilon q - lE) + T_a \tag{2.2}$$

where ε is emissivity, q is net infrared radiation, and l is the latent heat of vaporization of water (= 2500 kJ/kg).

The relationship between solar transmittance τ and MRT reduction by adaptation measures, such as an awning, is shown in Fig. 2.10. If the influence of long-wave radiation was ignored, complete shielding of solar radiation decreased the MRT by 15°C.

The relationship between the surface temperature T_s of the adaptation measures and the solar absorptance a when the heat transfer coefficient h is 23 $W/(m^2K)$, emissivity ε is 0.97, and net infrared radiation q is -93 W/m^2 for different values of the evaporation rate E is shown in Fig. 2.11. Although net infrared radiation q and the evaporation rate E varied depending on weather conditions such as surface temperature, air temperature, and wind velocity, they were set to specific values to allow simple evaluation. Even if the evaporation rate E was 0 $L/(m^2h)$, when the solar radiation absorptance a was 0.1, the surface temperature T_s was almost the same as the air temperature. The surface temperature T_s when the heat transfer coefficient h is 46 or 92 $W/(m^2K)$ is shown in Fig. 2.12. A fractal-shaped sunshade was developed focusing on the utilization of the effect caused by increasing the heat transfer coefficient [5]. As the heat transfer coefficient h increased, the surface temperature T_s approached the air temperature value regardless of the solar absorptance a and evaporation rate E.

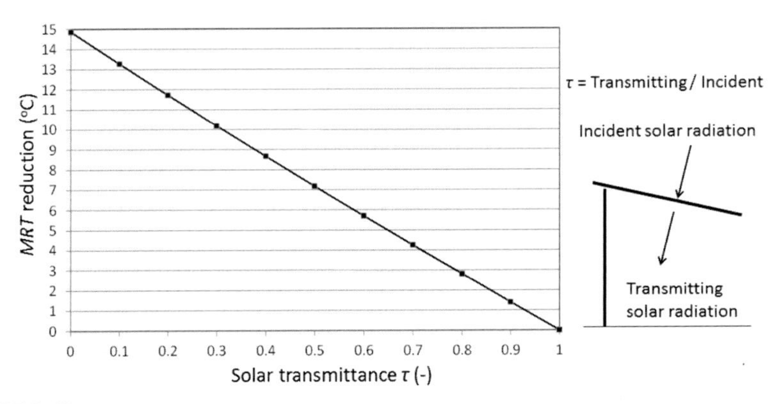

FIGURE 2.10

Relationship between solar transmittance τ and mean radiant temperature (MRT) reduction by the adaptation measures.

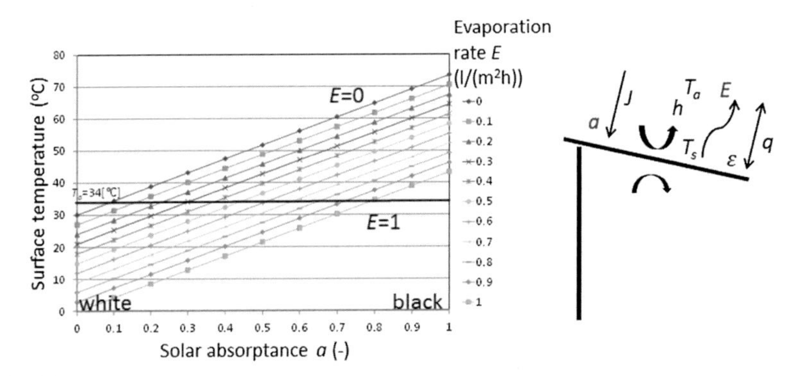

FIGURE 2.11

Relationship between the surface temperature T_s of the adaptation measures and the solar absorptance a when the heat transfer coefficient h is 23 W/(m^{2}K), emissivity ε is 0.97, and net infrared radiation q is -93 W/m^2 for different values of the evaporation rate E.

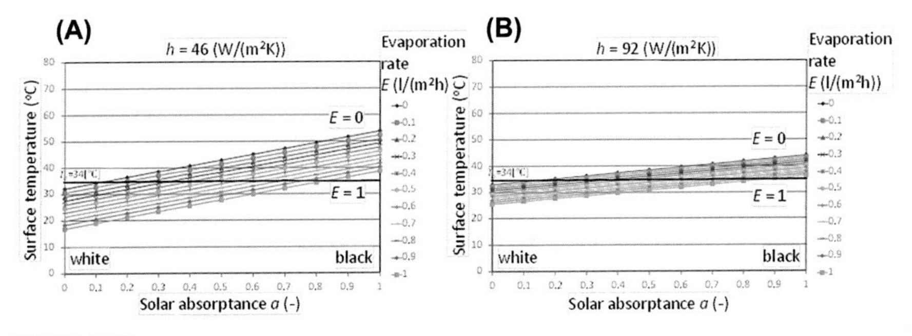

FIGURE 2.12

Surface temperature T_S when the heat transfer coefficient h is 46 (A), h is 92 W/(m^{2}K) (B).

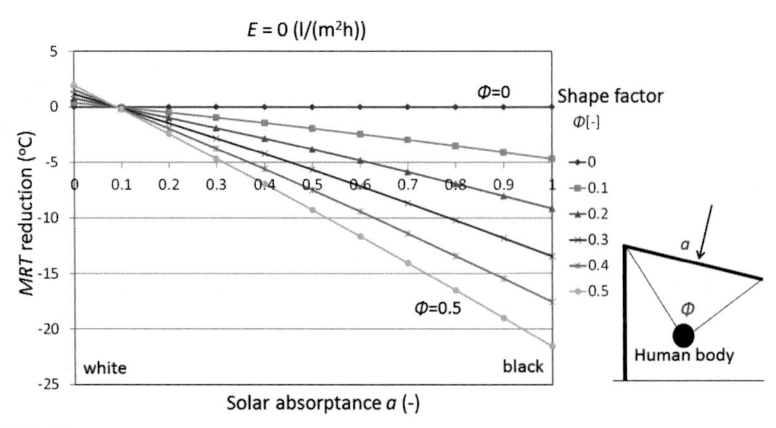

FIGURE 2.13

Relationship between the mean radiant temperature (MRT) reduction and the solar absorptance a when the evaporation rate E is 0 L/(m^{2}h) for different values of the shape factor Φ of the human body.

The relationship between the MRT reduction and the solar absorptance a when the evaporation rate E is 0 and 1.0 L/(m^{2}h) for different values of the shape factor Φ of the human body is shown in Figs. 2.13 and 2.14. If no evaporation occurs, when the shape factor Φ and solar absorptance a were large, the MRT increased due to the effect of long-wave radiation from the adaptation measures. Conversely, if large evaporation occurs, when the shape factor Φ was large and solar absorptance a was small, the MRT decreased due to the effect of long-wave radiation from the adaptation measures. These two figures show the effects on MRT by high- and low-temperature object above the human body.

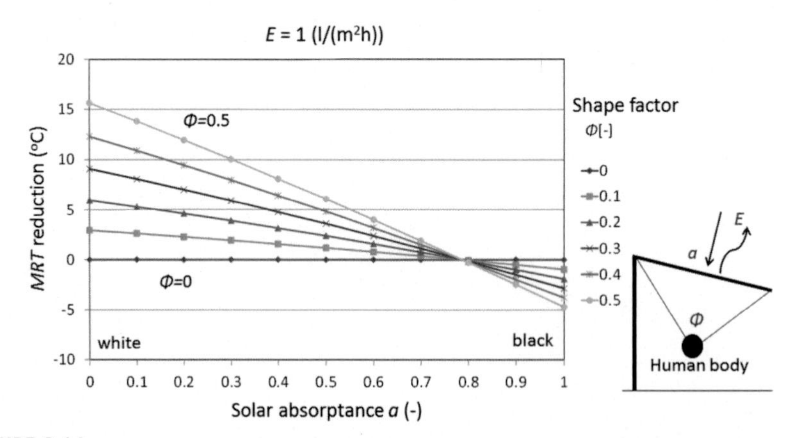

FIGURE 2.14

Relationship between the mean radiant temperature (MRT) reduction and the solar absorptance a when the evaporation rate E is 1.0 L/(m^{2}h) for different values of the shape factor Φ of the human body.

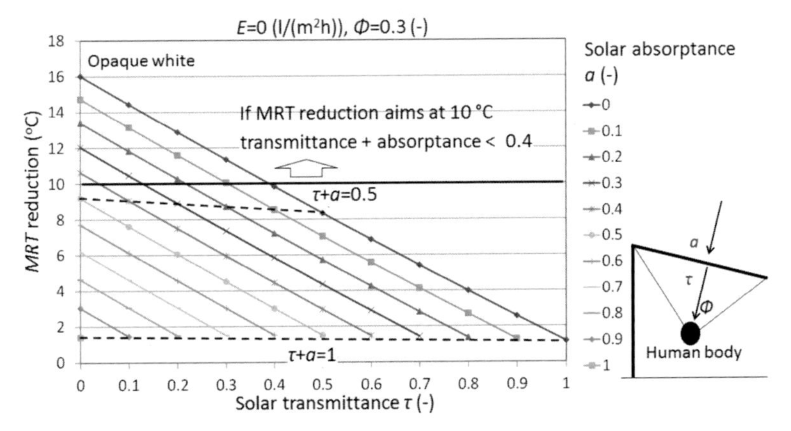

FIGURE 2.15

Relationship between the mean radiant temperature (MRT) reduction and the solar transmittance τ when the evaporation rate E is 0 L/(m^{2}h) and the shape factor of the human body Φ is 0.3 for different values of the solar absorptance a.

The relationship between the MRT reduction and the solar transmittance τ when the evaporation rate E is 0 L/(m^{2}h) and the shape factor of the human body Φ is 0.3 for different values of the solar absorptance a is shown in Fig. 2.15. When the targeted MRT reduction was 10°C, the required solar transmittance τ plus solar absorptance a was 0.4 or less.

The relationship between the MRT reduction and the solar transmittance τ when the evaporation rate E is 1.0 L/(m^{2}h) and the shape factor of the human body Φ is 0.3 for different values of the solar absorptance a is shown in Fig. 2.16. If the evaporation rate E was 1.0 L/(m^{2}h) or more, MRT decreased by 10°C regardless of solar transmittance τ and solar absorptance a.

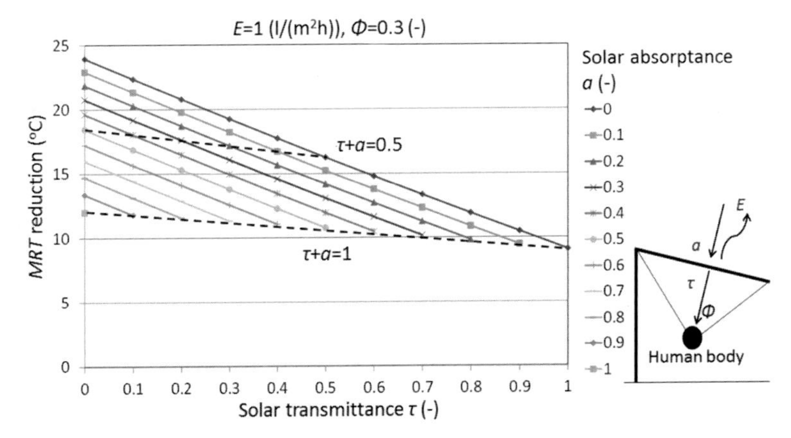

FIGURE 2.16

Relationship between the mean radiant temperature (MRT) reduction and the solar transmittance τ when the evaporation rate E is 1.0 L/(m^{2}h) and the shape factor of the human body Φ is 0.3 for different values of the solar absorptance a.

3. Example of field experiment results

To understand adaptation to thermal environment, we will provide examples of the effects of the differences of thermal environment on human physiological and psychological responses and space utilization.

3.1 Evaluation of the effects of the measures taken against the thermal environment on human psychological responses [31]

As a technological countermeasure to thermal environment, we devised and evaluated the thermal environment mitigation effect using a tent-like rescue facility made of a nonwoven fabric, carrying a photocatalyst. To mitigate the urban thermal environment, a tent made of hydrophilic nonwoven fabric treated with a photocatalyst was developed. The experiments were carried out to evaluate the evaporative performance of the fabric and the cooling effects of the tent made out of it. In this, as shown in Fig. 2.17, a nonwoven fabric with hydrophilicity was attached to the side of a tent by a photocatalyst, and water was sprayed, expecting a rise in temperature on the vertical surface through evaporative cooling. This tent was setup during an outdoor event in the summer and was evaluated based on the thermal environment and psychological reporting surveys.

Fig. 2.18 shows the changes in SET* and the MRT, indicating the measurement results of the thermal environment. The SET* inside the tent, which has taken measures against the thermal environment, was 3–5°C lower than the outside; hence, it was confirmed that thermal comfort was improved. In particular, the significant difference in the MRT suggested that solar radiation shielding and vertical surface cooling greatly improved the radiation environment. Fig. 2.19 illustrates the relationship between SET*, thermal sensation, and comfort from the results of the psychological declaration survey, which showed a positive correlation between SET* and thermal sensation and a negative correlation between SET* and thermal comfort.

It can be gathered from the results that the tent could be a cool and comfortable space, and psychological reactions confirmed the improvement in the thermal environment.

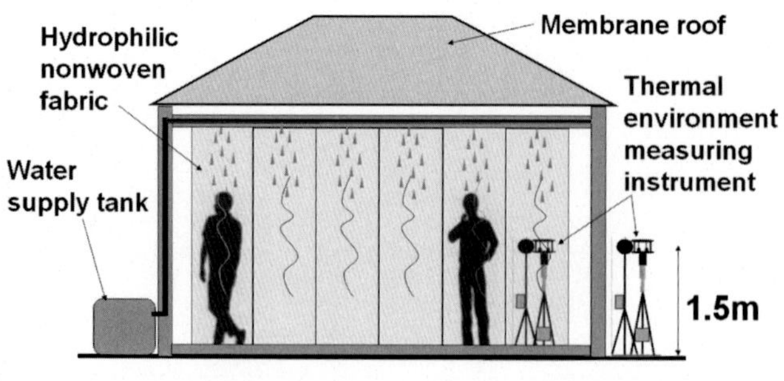

FIGURE 2.17

Tent unit using nonwoven fabric with hydrophilicity.

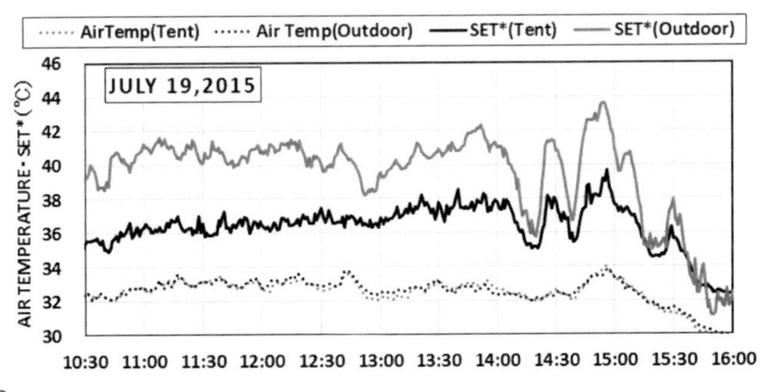

FIGURE 2.18

Time change of new standard effective temperature (SET*) and mean radiant temperature (MRT).

3.2 Evaluation of the measures taken against the thermal environment on human physiological responses [32]

As a new countermeasure technology for thermal environment, a water-cooled bench was devised, and its effects were evaluated through experiments. This bench cooled the seat surface by allowing water to flow from the side into an aluminum bench with a hollow seat core material that can suppress the rise in temperature, compared to a standard bench made of artificial wood. The countermeasures for heat by water-cooled benches are different from environmental changes, making an evaluation of the heat balance of the human body necessary.

Therefore, human physiology experiments were conducted on the subjects wherein the temperature of the supplied water was adjusted so that the seating surface temperature was about 28°C. The four conditions for the subjects sitting

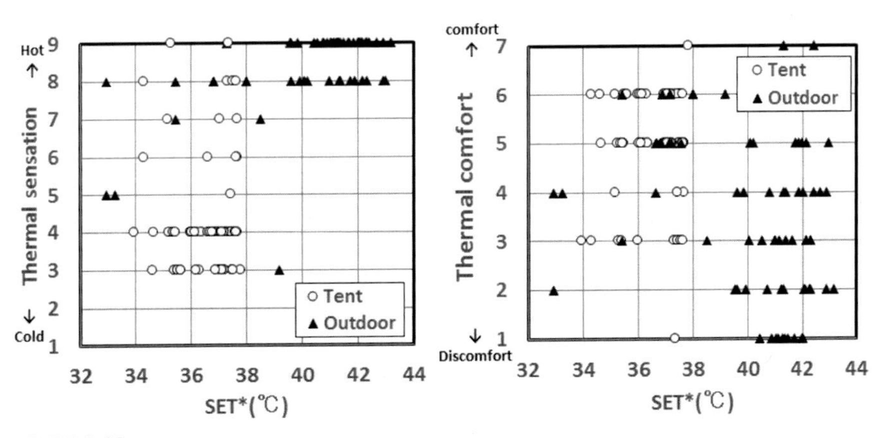

FIGURE 2.19

Relationship between new standard effective temperature (SET*) and psychological declaration.

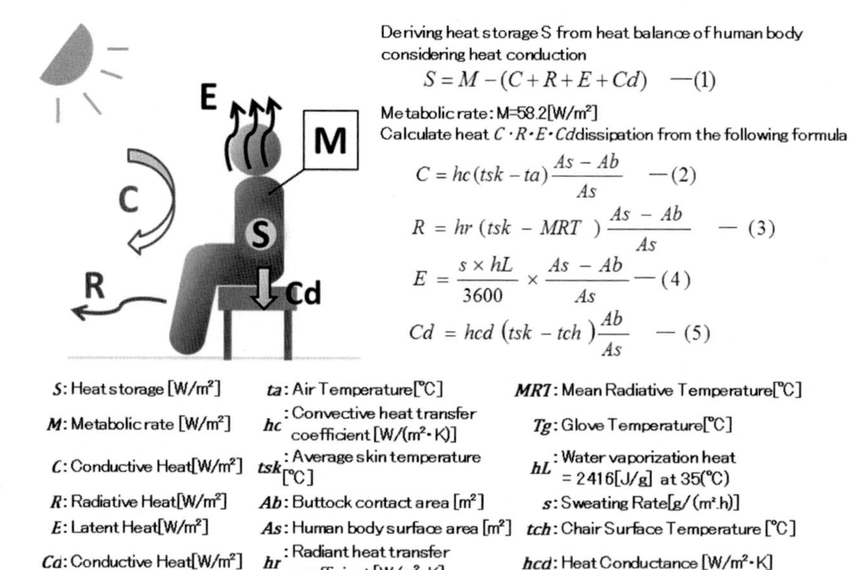

FIGURE 2.20

Conceptual diagram of heat balance of human body and calculation method.

outdoors were sitting on an insulated chair in the sun, tent, and heat-treating facility, and sitting on a water-cooled bench at the heat-treating facility. Each of them was seated for 15 min. The heat load on their bodies was calculated using the heat balance conceptual diagram. Fig. 2.20 shows the calculation method. Fig. 2.21 shows the calculation results of the heat balance for each subject. The results revealed that the radiant heat radiation was lesser in the tents and countermeasure facilities, compared to the sun, despite individual differences. Furthermore, the amount of

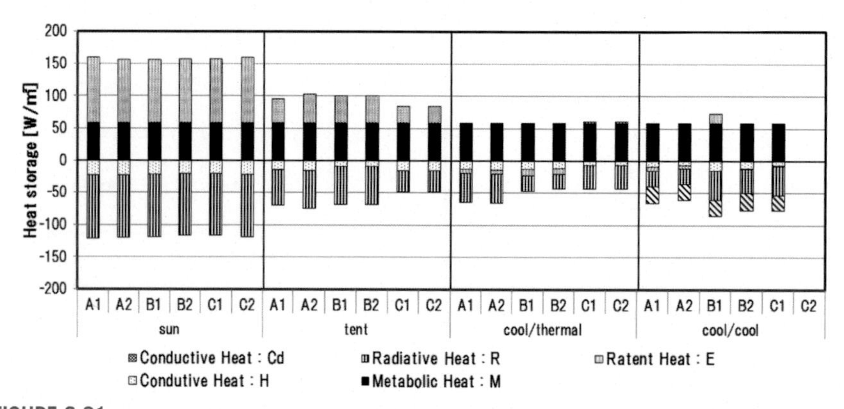

FIGURE 2.21

Calculation results of heat balance by each subject.

heat released from the human body on being seated on the water-cooled bench was more than that on the heat-insulated chair. The heat conducted on the water-cooled bench was high.

Hence, analyzing the heat balance of the human body helps evaluate the effect of thermal stress reduction, which promotes heat dissipation from the body, using a water-cooled bench.

3.3 Evaluation of the effects of measures taken against the thermal environment on space utilization [33]

If countermeasures against heat are taken as an adaptation measure and a comfortable environment can be created in a thermal environment, people can actively use outdoor spaces even in a thermal environment.

Therefore, we investigated the relationship between the thermal environment and space utilization by people. The survey was conducted in a thermal environment with residents on four benches in an outdoor plaza. Two of the four benches were used as measures against the thermal environment (Fig. 2.22), and the remaining two were each located under the sun and the shade.

Fig. 2.23 shows the changes in the SET* indicating the measurement results of the thermal environment. Compared to the sun, the location of the benches as countermeasures and the one in the shade had a lower SET*, confirming the effect of thermal countermeasures.

As the numbers of bench users vary depending on the time of the day, it was difficult to capture trends common to all time zones. Therefore, a standardization was performed, and the number of standardized users was calculated. The standardized number of users is the ratio of the average number of users in a certain place/time zone. Furthermore, the probability density distribution and the probability function were approximated from the relationship between the obtained SET* and the number of standardized users. The calculation results are shown in Fig. 2.24. It can be seen that this space was easy for humans to use when the SET* fell below 30°C. According to the results at 12:00 on the survey day, the SET* was reduced by about 5°C

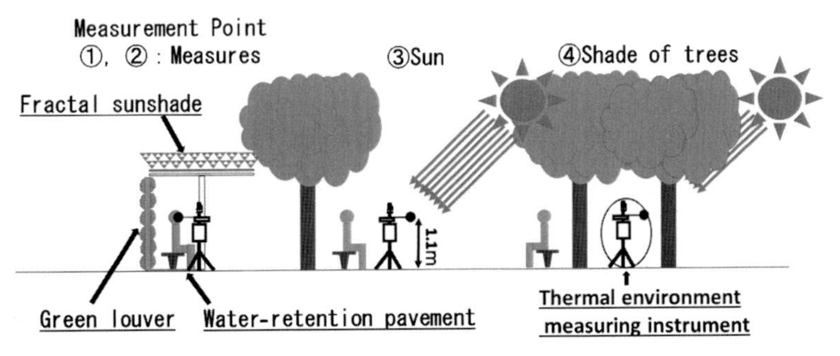

FIGURE 2.22

Outline of the survey site.

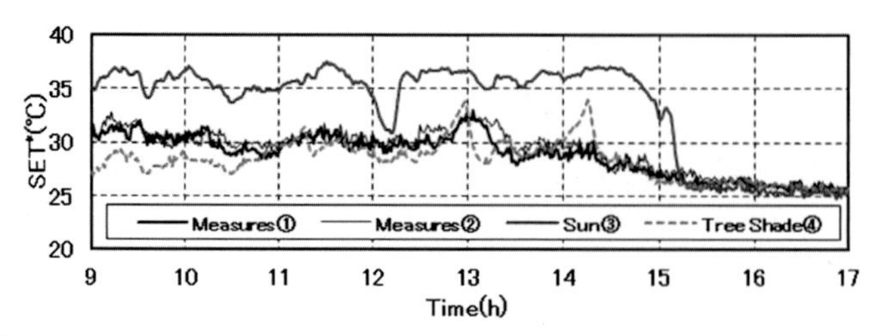

FIGURE 2.23

Time change of new standard effective temperature (SET*).

from 35.0°C in the sun by implementing measures against the heat, increasing the number of standardized users by 0.76. Considering the average number of users at 12:00, an increase of 1.40 users per minute could be expected.

From the results, the relationship between the thermal environment and the number of users indicates the possibility of promoting the study and implementation of heat countermeasures for the purpose of promoting the use of space.

3.4 Summary and future development

In addition to the "mitigation measures" that alleviate the factors occurring in the past, the idea of "adaptation measures" has been proposed as a countermeasure against temperature rise caused by climate change and the heat island phenomenon. In the field of adaptation to thermal environment, it is necessary to take measures against heat in the city to reduce human thermal stress. Furthermore, for people to actively use outdoor spaces, it is important to create a space that is a comfortable thermal environment. Therefore, a survey was conducted on the thermal environment, the physiological and psychological reactions of the human body, and the use of outdoor spaces by people. The relationship between the thermal environment

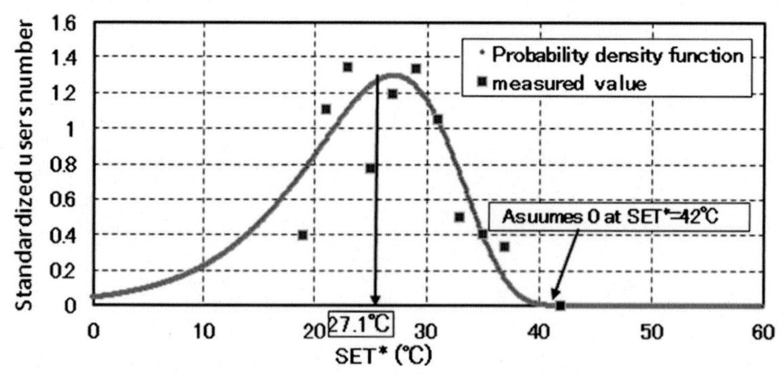

FIGURE 2.24

Relationship between new standard effective temperature (SET*) and probability density distribution of standardized users.

and human physiology, psychological reaction, and space use suggested the existence of thresholds for outdoor thermal environment and thermal comfort. In the future, it will be important to set up a permissible range of thermal environment befitting outdoor space use, and develop a method for designing the thermal environment within the permissible range.

4. Examples of adaptation measures at construction sites

4.1 Thermal conditions at construction sites

According to statistics from the Japanese Ministry of Health, Labor and Welfare regarding the number of patients affected by heat-related illness, in the decade between 2009 and 2018, approximately 500 people suffered heat stroke while performing work activities in Japan. Nearly half of these patients were construction and manufacturing workers [34]. Both the number of deaths and the number of patients diagnosed with a heat-related illness are higher in the construction industry than in any other industry.

Traditional prevention measures to avoid heat disorder at a construction site are as follows. Supervisors measure WBGT using a portable sensor or obtain WBGT guideline information from the web (e.g., Heat Illness Prevention Information provided by the Japanese Ministry of Environment [35]) and post it in a common area at the worksite. Next, they monitor workers and use WBGT information to instruct them about managing their water and salt intake. However, owing to insufficient WBGT alert notifications, most workers are unknowingly exposed to different WBGT environments at various places in the construction site. In particular, workplaces in the summer are often hotter than the location where WBGT is measured.

Fig. 2.25 shows the distribution of difference between temperature experimentally obtained at various places in different construction sites and that measured at government meteorological stations in multiple regions of Japan. Positive values on the horizontal axis show where WBGT in a construction site is higher than temperatures measured concurrently at meteorological stations. Because 81.3% of site data are higher than meteorological station data, collecting a large amount of WBGT data at various places in the construction site and using it to evaluate work conditions can be very effective in decreasing the risk of heat disorder in workers.

4.2 IoT-based health monitoring system for preventing workers' heat disorder

Recent rapid advances in human sensing technologies facilitate real-time monitoring of heart rate (HR) during work activities. Several types of wearable sensors, such as a wristband, shirt, etc., have been applied to prevent heat disorder at construction sites. Obayashi Corporation, a Japanese construction company, built an IoT-based system for monitoring workers' health conditions, such as the potential for heat

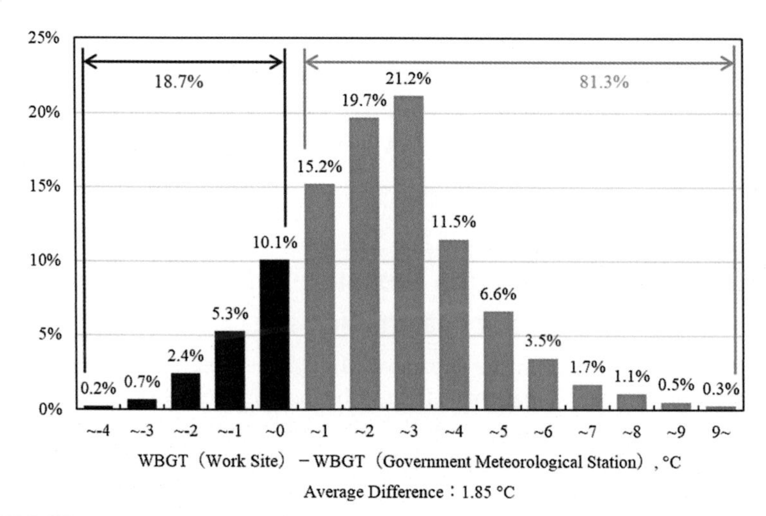

FIGURE 2.25

Distribution of difference between temperature experimentally obtained with wet-bulb globe temperature (WBGT) at various places in different construction sites and that measured at government meteorological stations in multiple regions of Japan.

disorder, and has operated the system for several years using both a shirt- and wristband-type HR sensor. The latest sensor model is a wristband type, as shown in Photo 2.1, which transmits multi-intensity beacon signals to detect the workers' position through IoT gateways installed at the construction site. The system is operated on a cloud-based service in which all HR data collected by wristband sensors and all environmental data collected by the originally developed WBGT sensors

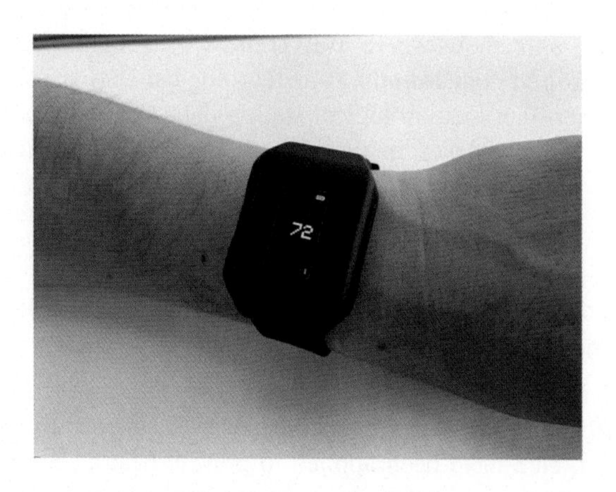

PHOTO 2.1

Wristband sensor for IoT-based health monitoring system.

[36] installed at the construction site are archived and analyzed to calculate each worker's HR threshold for triggering an email alert.

Fig. 2.26 shows a schematic diagram of IoT-based health monitoring system for construction workers which can evaluate risk potential for heat disorder. Fig. 2.27 shows a list of conditions which indicate an alert situation calculated using workers' HR and surrounding environment WBGT data.

There is no standard HR threshold value for officially determining heat disorder. A small number of guidelines referencing HR threshold to suspend work under hot conditions are described in ISO 9886 [37] and ACGIH—Heat Stress and Strain TLV [38]. According to these standards, HR during work should be lower than the value predicted by Eq. (2.3).

$$HR = 185 - 0.65 \times age \tag{2.3}$$

The sustained (several minutes) HR should be lower than the value predicted by Eq. (2.4).

$$HR_{sustained} = 180 - age \tag{2.4}$$

The IoT-based health monitoring system uses these equations provisionally as part of a logic to calculate the current HR threshold. An accumulation of data using this system will lead to more reliable logic calculations in the near future.

WBGT is generally used to prevent heat disorder during construction work. Because WBGT is not uniform throughout a construction site, as mentioned previously, actual WBGT values measured in the workers' environment should be collected and used as a threshold. Because the developed WBGT sensor is cheaper

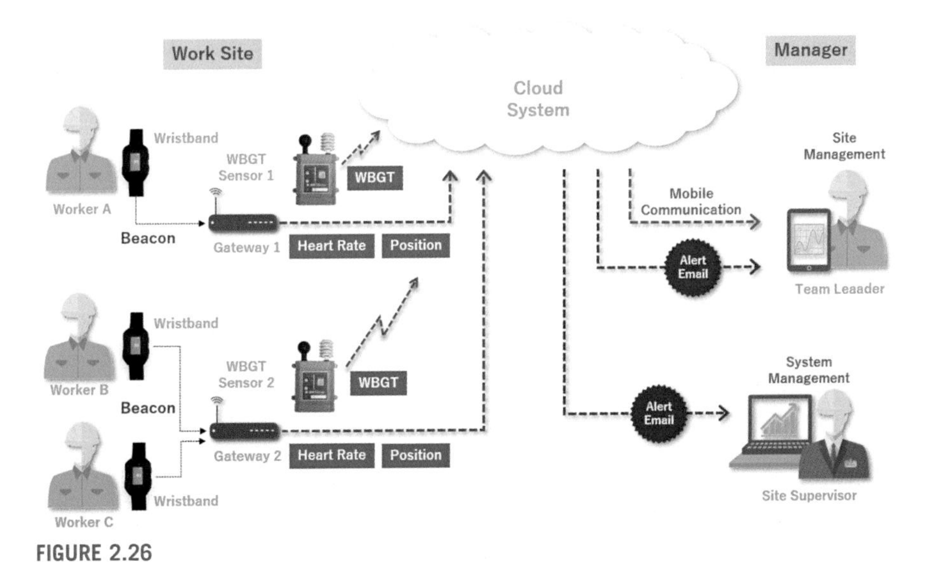

FIGURE 2.26

Schematic diagram of IoT-based health monitoring system for construction workers.

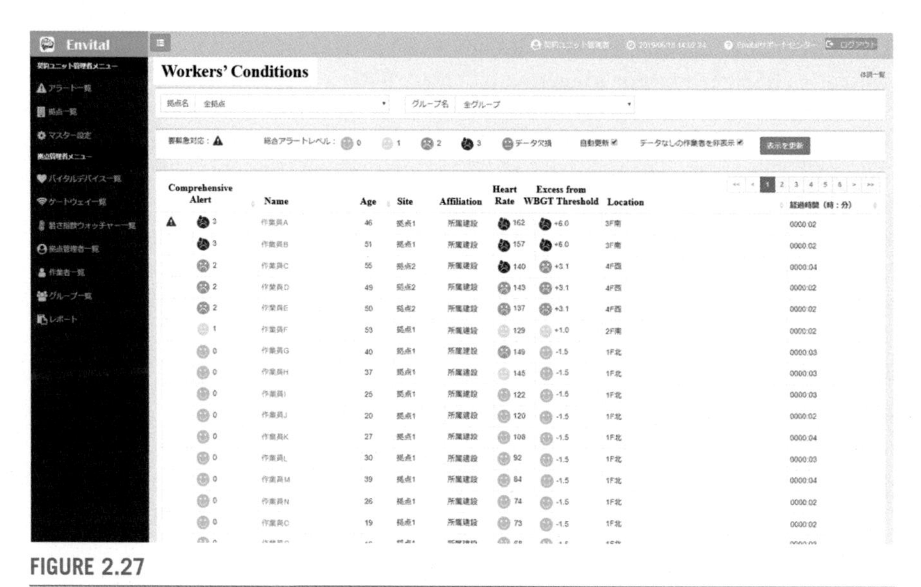

FIGURE 2.27

List of conditions indicating an alert status calculated using workers' heart rate (HR) and surrounding environment wet-bulb globe temperature (WBGT) data.

than those on the market and has a multihop radio communication system, many sensors (e.g., one or two per floor) can be distributed all over the sites, and each worker's environment can be considered in the system. The system includes settings for metabolic rate, amount of clothing, acclimation to high temperatures, and the presence of air current used to calculate more accurate WBGT, based on the WBGT standard (ISO 7243).

Alert information based on the logic appears on the PC screen and mobile devices, as this is a cloud-based system. Alert emails can also be sent to supervisors or workers themselves at their request. Photo 2.2 shows a supervisor asking a worker about his health condition during work after the supervisor received an alert email.

4.3 Expected effects

Most workers safely worked under hot conditions above the WBGT threshold. Although a few abnormal HR variations (e.g., extremely high) were discovered in the large quantity of collected data, it has not yet been concluded that these variations are related to a symptom of heat disorder. Some tendencies associated with HR characteristics were found in daily HR and WBGT alert charts. Fig. 2.28 shows the table of WBGT (left column) and HR (right column) alert levels collected before and after the 1-week summer holiday of a 59-year-old worker. Prior to the holiday, his HR was not high despite a high WBGT. By contrast, his HR tended to be high just after the holiday and dropped gradually despite the fact that WBGT was still

PHOTO 2.2

A supervisor asks a worker about his health condition during work after the supervisor received an alert email.

high. This phenomenon is due to the influence associated with the loss of acclimation to high temperatures.

Fig. 2.29A and B show HR time variations which were collected from three workers at two different construction sites. Site A is a typical architectural construction site; work was at the ground-level stage, and therefore, little shade was at the site. Site B is a factory renewal construction site where blast furnace facilities were operated, and its air temperature was extremely high. The data were collected during a 1-week period in the summer of 2016.

Fig. 2.30 shows the environmental conditions when the HR data were collected. The graphs indicate the time variations of the degree of excess from the WBGT threshold measured near the work area at each site. In site A, the degree of excess showed diurnal variation in accordance with air temperature and solar radiation, because most of the construction area was in the sun. Meanwhile at site B, the degree of excess indicated an almost constant temperature level at over 8°C above

	Excess from the WBGT Threshold (°C)											
	7-8h	8-9h	9-10h	10-11h	11-12h	12-13h	13-14h	14-15h	15-16h	16-17h	17-18h	18-19h
1-Aug		1.0	2.2	3.9	4.1	5.2	5.0	4.9	4.1	5.2	5.3	4.2
2-Aug	0.0	1.5	3.3	3.8	4.4	4.2	4.4	2.7	1.3	0.2	0.0	
3-Aug	0.0	0.0	0.2	2.5	4.7	5.6	5.4	5.5	5.1	5.0	5.6	
4-Aug	0.0	0.0	2.0	3.7	5.5	6.5	5.7	5.5	5.5	5.2	5.2	2.2
5-Aug	0.7	22.0	3.1	3.2	3.7	6.0	6.2	3.8	5.0	5.5	3.2	
7-Aug		0.0	0.0									
9-Aug	0.2	1.0	3.2	4.2	5.4	6.1	6.2	6.1	5.4	4.9	4.5	
Summer Holiday (7 days)												
17-Aug		0.7	0.2	1.2	5.0	6.0	6.7	6.7	6.5	6.2	4.9	
18-Aug		0.2	1.2	3.0	3.7	5.5	6.5	3.6	5.5	5.5	5.0	
19-Aug	0.0	0.0	1.0	3.2	3.2	3.7	3.2	3.7	3.9	3.2	2.7	
20-Aug	0.0	1.6	4.5	4.4	3.5	4.4	5.7	6.0	5.7	6.2	6.0	
21-Aug	0.0	0.7	4.2	5.0	5.0	9.2	8.5	6.2	6.5	6.0	3.7	
25-Aug	0.7	2.2	3.7	4.2	5.4	5.7	6.9	6.7	6.4	7.0	7.0	
28-Aug	0.0	0.0	0.0	1.0	3.0	3.0						
29-Aug	0.0	0.0	0.0	1.7	2.7	4.9	5.2	5.5	5.4	4.0	3.2	
31-Aug	0.0	0.0	0.7	1.1	3.2	4.5	4.2	2.9	1.4	1.1	0.0	0.0

	Comprehensive Alert Level (0 - 3)											
	7-8h	8-9h	9-10h	10-11h	11-12h	12-13h	13-14h	14-15h	15-16h	16-17h	17-18h	18-19h
1-Aug		0	0	0	0	0	1	0	0	0	0	0
2-Aug	0	1	0	0	0	1	0	0	0	0	0	
3-Aug	0	0	0	0	0	0	0	0	0	1	0	
4-Aug	0	0	0	1	1	0	2	0	1	0	2	0
5-Aug	0	1	0	0	0	0	1	0	0	1	0	
7-Aug	0	0	0									
9-Aug	0	1	1	1	0	0	3	2	0	1	0	
Summer Holiday (7 days)												
17-Aug	0	3	3	1	2	2	3	3	3	3	2	
18-Aug	0	1	0	1	2	1	2	2	2	2	0	0
19-Aug	0	0	0	1	1	0	1	1	2	2	1	
20-Aug	0	0	1	1	0	0	2	1	0	0	0	
21-Aug												
25-Aug	0	1	0	0	0	1	0	1	1	2	2	
28-Aug	0	0	0	1	1	1						
29-Aug	0	0	0	0	1	2	2	2	2	2	0	
31-Aug	0	0	1	0	0	0	1	0	0	0	0	0

FIGURE 2.28

Alert levels of wet-bulb globe temperature (WBGT) (left column) and heart rate (HR) (right column) collected before and after the 1-week summer holiday of a 59-year-old worker.

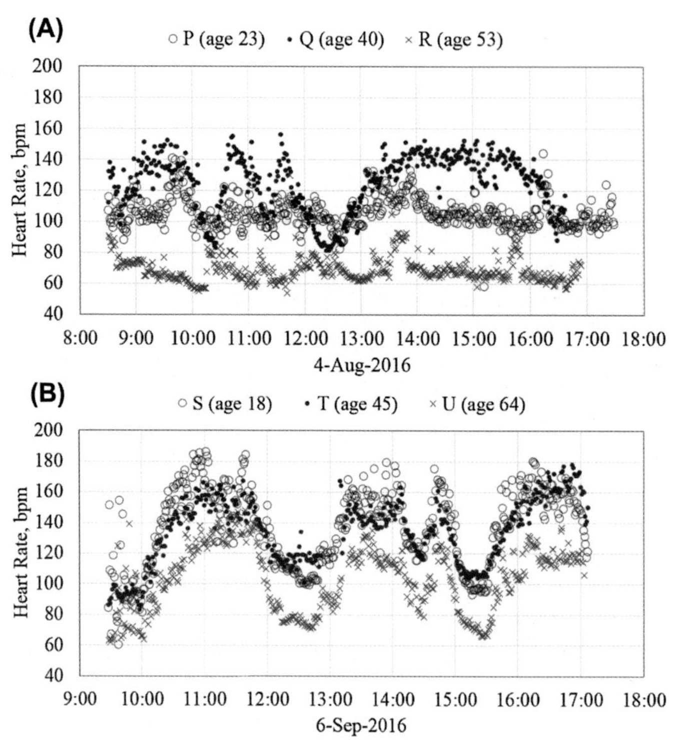

FIGURE 2.29

Time variations of heart rate (HR): (A) Site A—Typical architectural construction site. (B) Site B—Factory renewal construction site where blast furnace facilities were operated.

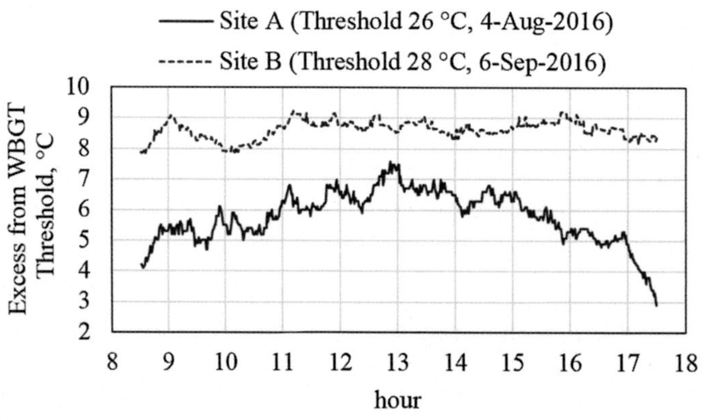

FIGURE 2.30

Time variations of the degree of excess from the wet-bulb globe temperature (WBGT) threshold at Sites A and B.

the WBGT threshold, which means that the workers experienced a condition of intense heat throughout the whole day. The HR variations of the three workers in site A show various patterns of HR fluctuation, as shown in Fig. 2.29A. Worker P (23 years old) had a relatively flat variation with a high HR level (100–120 bpm). Worker Q (40 years old) had a seesaw variation with a high HR level (80–150 bpm). Worker R (53 years old) had a relatively flat variation with a low HR level (60–80 bpm). The different HR variation patterns could be related to the workers' occupations and work duties on a particular day. Conversely, owing to high temperature conditions, all three workers in site B had a large HR variation regardless of their age, as shown in Fig. 2.29B. In particular, at several times during the day, workers S (18 years old) and T (45 years old) had an extremely high HR level at over 160 bpm. Even during break times (between 1200 and 1300, around 1400, and around 1500) their HR remained high at approximately 100 bpm. At both sites, veteran workers R and U (64 years old) had a comparatively lower HR than the other workers, which suggests that, based on experience, veteran workers are more knowledgeable about work efficiency and their physical limits.

Fig. 2.31A and B show the relationship between WBGT and HR for the workers at site A, classifying them into two groups: younger than and older than 40 years old, respectively. The maximum HR tended to go up as WBGT increased, particularly when the temperature exceeded approximately 30°C in both age groups. Workers older than 40 years tended to have high maximum HR under high WBGT conditions, caused by the influences of their age or workload. Conversely, some workers older than 40 years showed relatively low HR, which suggests that veteran workers control their workload as described in Fig. 2.29A and B. In other work sites, the middle-aged workers in their 40s and 50s tended to have high HR, which indicates that these age groups carry a heavy workload beyond their physical ability.

4.4 Expectations for the future

The combined use of the latest vital sensing and wireless communication technology at construction sites is rapidly progressing. If workers' vital data are continuously uploaded, we can provide safer work environments using the system described above. Because the application of IoT-based safety management at construction sites has just begun, the relationship between HR and health conditions or the surrounding environment is not yet clear. Continuous operation of this system at various construction sites will accumulate a large amount of data, which can gradually help clarify that relationship. As a result, the threshold of alert will evolve to become more appropriate for construction work activities. Because the variations of HR are influenced by many factors such as temperature, age,workload, type of work, etc., as described in this article, we consider that various types of big data should be collected at the actual construction site, not in the laboratory.

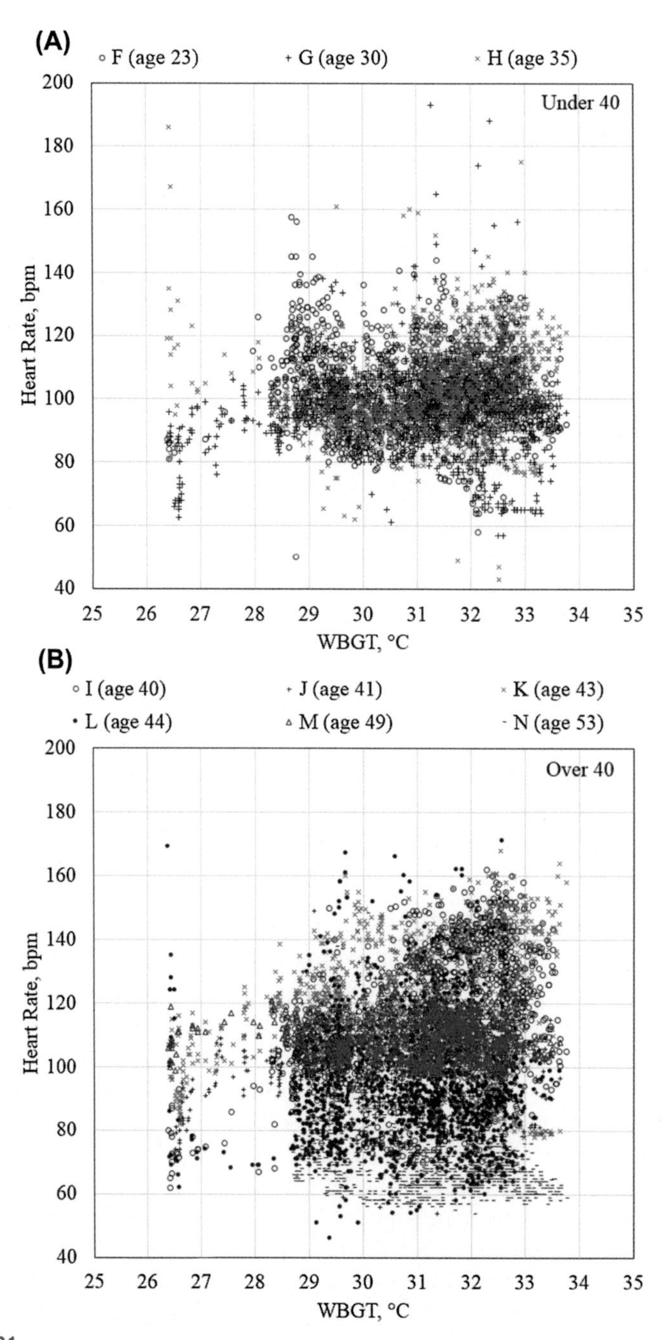

FIGURE 2.31

Relationship between wet-bulb globe temperature (WBGT) and heart rate (HR) for workers in site A: (A) Younger than 40 years; (B) older than 40 years.

5. Summary

Through several examples of the effects of adaptation measures obtained by demonstrative experiments, it can be seen shielding of solar radiation to pedestrians is a more effective method of lowering MRT and SET*, PET. The influence of the solar transmittance of adaptation measures such as shading, on MRT is approximately 1.5°C per 0.10. The influence of the solar absorptance of adaptation measures such as an awning, on MRT is approximately 1.0°C per 0.10, which also depends on the shape factor between the human body and adaptation measures. The influence of the evaporation rate on MRT is approximately 1.0°C per 0.10 L/(m^2h). If a shielding device that reflects a large amount of solar radiation and facilitates high levels of evaporation is developed, MRT and SET*, PET will both decrease.

Moreover, the evaluation results of the influence on human's physiological and psychological thermal environment were explained based on measured data. Then, examples of adaptation measures at construction site were also described with reference to specific data. The combined use of the latest vital sensing and wireless communication technology is explained by showing application examples at construction sites.

References

[1] The Ministry of the Environment of Japan. Heat countermeasure guideline in the city. 2018. http://www.env.go.jp/air/life/heat_island/guidelineH30/gudelineH30_all.pdf. [Accessed 18 December 2011].

[2] Center for Environmental Information Science. Report on consignment work of survey and verification for the creation of a low-carbon city using surplus groundwater etc.. Report entrusted by the Ministry of the Environment in 2016 fiscal year. 2017 [in Japanese].

[3] Osaka Heat Island Countermeasure Technology Consortium. Town planning idea competition considering urban heat island. 2015. http://osakahitec.com/result/index.html. [Accessed 18 December 2011].

[4] Takayama N, Yoshikoshi H, Yamamoto H, Iwaya K, Harada Y, Yamasaki T, Tateishi Y. Quantitative evaluation of mitigation effect for thermal load of solar radiation through the glass window by wall greening. Journal of Environmental Engineering 2011;661: 247–54 [in Japanese].

[5] Sakai S, Nakamura M, Furuya K, Amemura N, Onishi M, Iizawa I, Nakata J, Yamaji K, Asano R, Tamotsu K. Sierpinski's forest: new technology of cool roof with fractal shapes. Energy and Buildings 2012;55:28–34.

[6] Inoue T, Ichinose M, Nagahama T. Improvement of outdoor thermal radiation environment in urban areas using wavelength-selective retro-reflective film. In: PLEA; 2015. p. 48.

[7] Sakai H, Emura K, Igawa N, Iyota H. Reduction of reflected heat of the sun by retroreflective materials. Journal of Heat Island Institute International 2012;7–2:218–21.

[8] Takebayashi H, Moriyama M. Study on surface heat budget of various pavements for urban heat island mitigation. Advances in Materials Science and Engineering 2012:1–11.

[9] Akagawa H, Takebayashi H, Moriyama M. Experimental study on improvement of human thermal environment on a watered pavement and a highly reflective pavement. Journal of Environmental Engineering 2008;623:85–91 [in Japanese].
[10] Takebayashi H, Moriyama M. Study on the urban heat island mitigation effect achieved by converting to grass-covered parking. Solar Energy 2009;83(8):1211–23.
[11] Misaka I, Suzuki H, Mizutani A, Murano N, Tashiro Y. Evaluation of heat balance of wall greening. AIJ Journal of Technology and Design 2006;23:233–6 [in Japanese].
[12] Hirayama Y, Ohta I, Hoyano A. Development of a surface wetting passive cooling louver system with hydrophilic and water absorbing coating film and an evaluation of its fundamental performance by outdoor experiment. Journal of Heat Island Institute International 2015;10:24–34 [in Japanese].
[13] Yoon G, Yamada H, Okumiya M, Tsujimoto M. Study on cooling system by using dry mist, Validation of cooling effectiveness and CFD simulation. Journal of Environmental Engineering 2008;633:1313–20 [in Japanese].
[14] Farnham C, Nakao M, Nishioka M, Nabeshima M, Mizuno T. Study of mist-cooling for semi-enclosed spaces in Osaka, Japan. Procedia Environmental Sciences 2011;4:228–38.
[15] Kojima I, Yoshinaga M. Analysis of the effect by the material and color of awnings – discussion about the outdoor test method of SC-value based on JIS A 1422. In: Summaries of technical papers of annual meeting AIJ. D-2; 2013. p. 145–6 [in Japanese].
[16] Nishimura N, Nomura T, Iyota H, Kimoto S. Novel water facilities for creation of comfortable urban micrometeorology. Solar Energy 1998;64:197–207.
[17] Nouri AS, Costa JP, Santamouris M, Matzarakis A. Approaches to outdoor thermal comfort thresholds through public space design: a review. Atmosphere 2018;9(3).
[18] d'Ambrosio Alfano FR, Olesen BW, Palella BI, Povl O. Fanger's impact ten years later. Energy and Buildings 2017;152:243–9.
[19] Höppe P. The physiological equivalent temperature – universal index for the biometeorological assessment of the thermal environment. International Journal of Biometeorology 1999;43:71–5.
[20] Mayer H, Höppe P. Thermal comfort of man in different urban environments. Theoretical and Applied Climatology 1987;38:43–9.
[21] ACGIH. Threshold limit values for chemical substances and physical agents and biological exposures indices. Cincinnati, OH: American Conference of Governmental Industrial Hygienists; 2011.
[22] d'Ambrosio Alfano FR, Malchaire J, Palella BI, Riccio G. The WBGT index revisited after 60 years of use. Annals of Occupational Hygiene 2014;58(8):955–70.
[23] ISO 2017. Ergonomics of the thermal environment – assessment of heat stress using the WBGT (wet bulb globe temperature) index - ISO Standard 7243. International Organization for Standardization, Geneva.
[24] Gagge A, Fobelets P, Bergland L. A standard predictive index of human response to thermal environment. ASHRAE Transactions 1986;92:709–31.
[25] Ishii A, Katayama T, Shiotsuki Y, Yoshimizu H, Abe Y. Experimental study on comfort sensation of people in the outdoor environment. Journal of Architectural and Planning Research 1988;386:28–37 [in Japanese].
[26] Nakano J, Tanabe S. Thermal comfort and adaptation in semi-outdoor environments. ASHRAE Transactions 2004;110:543–53.
[27] Nagano K, Horikoshi T. New index indicating the universal and separate effects on human comfort under outdoor and non-uniform thermal conditions. Energy and Buildings 2011;43:1694–701.

[28] Watanabe S, Nagano K, Ishii J, Horikoshi T. Evaluation of outdoor thermal comfort in sunlight, building shade, and pergola shade during summer in a humid subtropical region. Building and Environment 2014;82:556–65.
[29] VDI 3787 Part 2. Environmental meteorology methods for the human biometeorological evaluation of climate and air quality for urban and regional planning at regional level part I: climate. 1998.
[30] Watanabe S, Horikoshi T, Ishii J, Tomita A. The measurement of the solar absorptance of the clothed human body – the case of Japanese, college-aged male subjects. Building and Environment 2013;59:492–500.
[31] Ikusei M, Yasuko M, Takuya S, Yu-ichi O, Hideki K, Yasushi I, Yasuyo H. Study on the mitigation effects of tent unit using non-woven fabric with hydrophilicity on thermal environment in urban area. In: Papers on environmental information science, vol. 30; 2016. p. 31–6.
[32] Ikusei M, Ken-ichi N, Yasushi I, Yasuyo H. Study on evaluation of thermal environment mitigation effect of water-cooled bench. In: 41st symposium on human-environment system, HES41 in Ueda; 2017.
[33] Ikusei M, Yasushi I, Yasuyo H, Ken-ichi N. Study on the effective utilization of outdoor space by mitigating thermal environment. In: Papers on environmental information science, vol. 31; 2017. p. 131–6.
[34] Ministry of Health, Labor and Welfare, Japan. Campaign for stopping heat disorder in 2019 (in Japanese), https://www.mhlw.go.jp/stf/newpage_03739.html. [Accessed 8 January 2019].
[35] Ministry of Environment, Japan. Heat illness prevention information. Available from: http://www.wbgt.env.go.jp/en/. [Accessed 8 January 2019].
[36] Obayashi Corporation News Release. Atsusa-shisu watcher. 2018. https://www.obayashi.co.jp/news/detail/news20180807_4.html. [Accessed 8 January 2019].
[37] ISO 9886. Ergonomics of the thermal environment: evaluation of thermal strain by physiological measurements. 2004. Geneva.
[38] ACGIH. Heat stress and strain TLV. In: American conference of government industrial hygienists. Cincinnati; 2012.

CHAPTER

Priority introduction place "hot spot" of adaptation measures

3

Hideki Takebayashi, Dr.

Associate Professor, Department of Architecture, Kobe University, Kobe, Japan

Chapter outline

1. Introduction

In order to mitigate the negative impact of urban heat islands, different strategies have been developed [1], such as solar radiation shade, urban ventilation, and mist spray, among others. The appropriate strategy should be applied depending on the characteristics of each location. It follows that the urban climate map is an effective tool for the identification of places that need intervention and, at the same time, for the evaluation of which adaptation technique should be applied at each location. Many of the existing studies analyze urban climate maps at urban scale and are focused on air temperature and wind distribution in the entire urban

Adaptation Measures for Urban Heat Islands. https://doi.org/10.1016/B978-0-12-817624-5.00003-8

area [2,3]. Differently, in order to focus on the radiation and wind effects on pedestrians, the analyzed area is at a district scale. Spatial distribution of air temperature and humidity are little for human thermal environment in a street canyon. Effects on wind and radiant environment due to building and urban block characteristics have been clarified in order to produce reliable urban climate maps at district scale, by using GIS building data instead of detailed calculation.

2. Climate characteristics in mesoscale: the urban and the surrounding area

2.1 Outline of calculation using the mesoscale weather research and forecasting model

As the major Japanese cities are located near the coast, they have climate characteristics affected by the sea breeze during fine summer days. The objective study areas, Tokyo, Osaka, and Nagoya which are the three major cities of Japan, are shown in Fig. 3.1. The outer square is domain 1 (3 km grid, 360 km square) and the inner filled square is domain 2 (1 km grid, 103 km square). Calculation condition is shown in Table 3.1. The period for which calculations were done was from August 1–31, 2010. Based on digital national land information (spatial resolution of 100 m) and a normalized vegetation index created from Landsat7 ETM + data, urban areas were classified into three categories according to the previous study [4]: high-rise and high-density, middle-rise and moderate-density, and low-rise and low-density. Present land use and potential natural vegetation in Tokyo, Osaka, and Nagoya are shown in

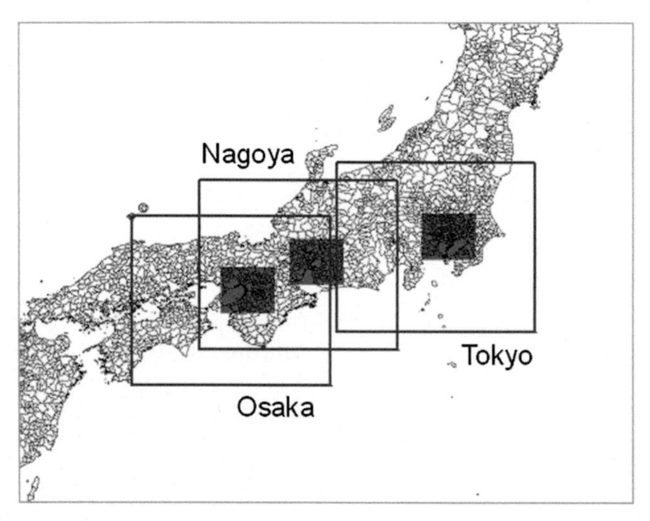

FIGURE 3.1

Objective study areas.

Table 3.1 Calculation condition for weather research and forecasting (WRF) model.

Calculation period	August 1–31, 2010
Vertical grid	28 layers from ground surface to 100 hPa
Horizontal grid	Domain 1: 3 km (120 * 120 grids), Domain 2: 1 km (103 * 103 grids)
Meteorological data	JMA: Mesoscale analysis (3 hourly, 10 km grids, 20 layers), NCEP: Final analysis (6 hourly, 1 degree grids, 17 layers)
Geographical data	Terrain height: Digital map (50 m resolution), Land use: Digital national land information (about 100 m resolution) + NVI (Landsat ETM+, November 5, 2001)
Microphysics process	Purdue Lin et al. Scheme
Radiation process	Long wave: RRTM long-wave scheme, Short wave: Dudhia short-wave scheme
Planetary boundary layer process	Mellor-Yamada-Janjic PBL scheme
Surface process	Urban area: Urban canopy model, Non-urban area: Noah LSM
Cumulus parameterization	None
Four-dimensional data assimilation	None

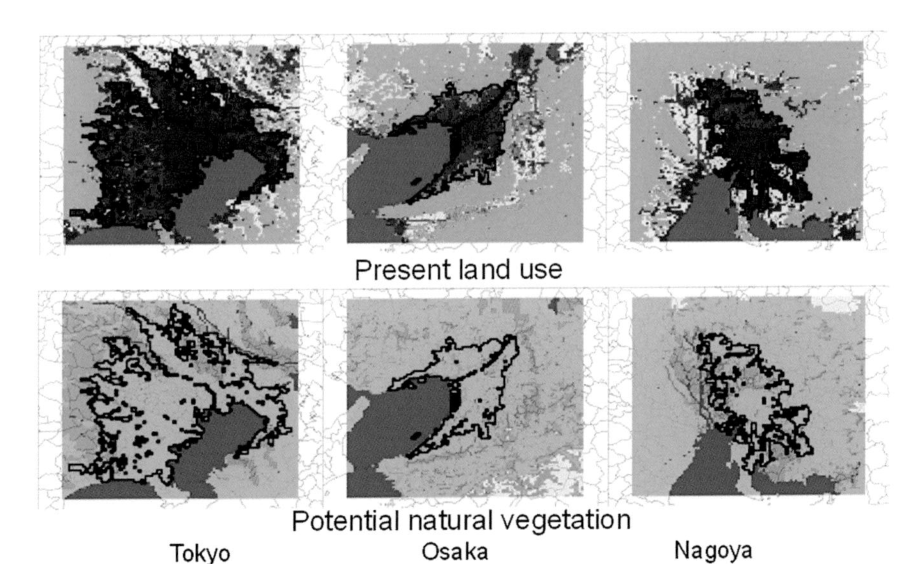

FIGURE 3.2

Present land use and potential natural vegetation in Tokyo, Osaka, and Nagoya.

Fig. 3.2. Potential natural vegetation is an ecological concept referring to the vegetation that would be expected given environmental constraints (climate, geomorphology, geology) without human intervention or a hazard event. The total number of these land use grids was 3698 in Tokyo, 1271 in Osaka, and 1416 in Nagoya. Frequency of urban land use in the three cities was larger in the coastal area and decreased gradually in the inland area. The number of urban land use grids along the coastal area in Nagoya was slightly smaller compared to that for Tokyo and Osaka.

2.2 Calculation results

The calculation results for air temperature at 2 m high under present land use conditions in Tokyo, Osaka, and Nagoya at 15:00 on August 26, 2010 are shown in the upper part of Fig. 3.3. In the three cities, air temperature was lower in the coastal areas and higher in the inland areas. The number of higher air temperature points in Tokyo was larger than that in the other cities, which are located mainly in the inland areas.

The calculation results for air temperature at 2 m high under natural vegetation land use conditions in Tokyo, Osaka, and Nagoya at 15:00 on August 26, 2010 are shown in the lower part of Fig. 3.3. The air temperature differences between current urban land uses and natural vegetation means urban heat island phenomena are caused by urbanization. Characteristics of air temperature distribution in natural vegetation land use are the same as those in urban land use, and air temperature for natural vegetation land use is 1–2°C lower than it is in present land use. Characteristics of air temperature distribution are recognized regardless of the presence

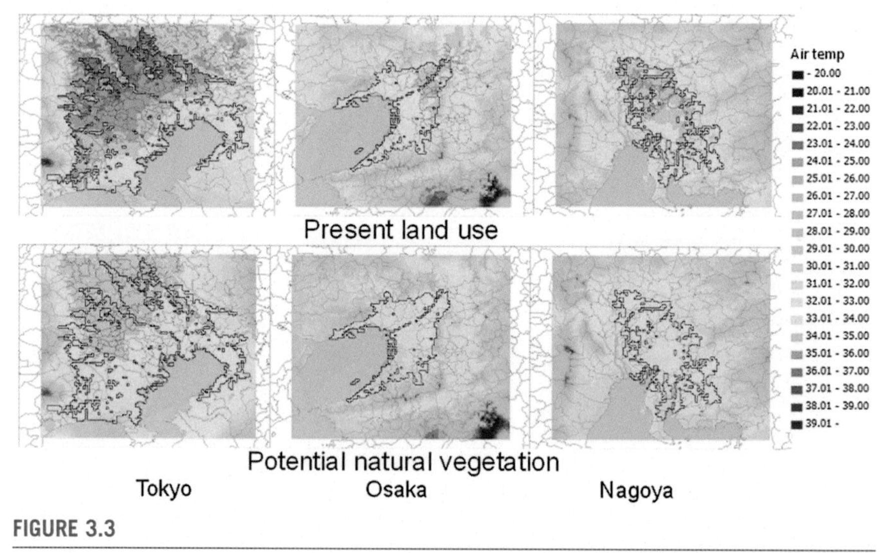

FIGURE 3.3

Calculation results for air temperature at 2 m high in Tokyo, Osaka, and Nagoya at 15:00 on August 26, 2010.

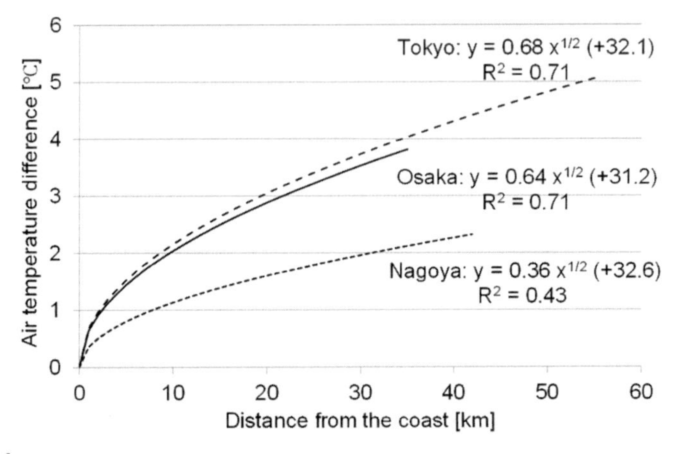

FIGURE 3.4

Relationship between distance from the coast and heat island intensity; approximation by 1/2 power equation.

of an urban area. A number of high-temperature points exist because urban areas spread inland where air temperatures tend to be relatively higher in Tokyo.

2.3 Relationship between distance from the coast and heat island intensity, thermal sensation indices

The relationship between distance from the coast and heat island intensity is shown in Fig. 3.4. According to the general urban boundary layer theory, heat island intensity is proportional to the square root of the distance from the urban boundary (the coast). The approximate curve for Tokyo overlaps the curve for Osaka but does not overlap the curve for Nagoya. The reason for this difference is that less urban land use occurs along the coastal area in Nagoya compared to the coastal areas of Tokyo and Osaka. The urban boundary layer is not very developed due to less urban land use along the coastal area of Nagoya, in comparison to Tokyo and Osaka. Inland air temperatures are about 3 degrees in Osaka and about 4 degrees in Tokyo higher than coastal areas.

Enthalpy, wet-bulb globe temperature (WBGT), new standard effective temperature (SET*), and physiological effective temperature (PET) distribution at 2 m high in Tokyo, Osaka, and Nagoya at 14:00 averaged on fine summer sea breeze day are shown in Figs. 3.5–3.8. Humidity, air temperature, wind velocity, enthalpy, SET*, WBGT, and PET at coastal and inland sites and their differences at 14:00 averaged in fine sea breeze days in Tokyo, Osaka, and Nagoya area is shown in Table 3.2. When the sea breeze blows into the urban area, absolute humidity increases as air temperature decreases. In that case, the cooling load on building air conditioning and outdoor human thermal sensation may not be mitigated due to sea breeze. It is evident from Fig. 3.5 that the enthalpy near the coast is slightly larger than that in the inland

Table 3.2 Humidity, air temperature, wind velocity, enthalpy, SET*, WBGT, PET at coastal and inland sites and their differences at 14:00 averaged in fine sea breeze days in Tokyo, Osaka, and Nagoya area.

	Humidity (g/kg)			Air temperature (°C)			Wind velocity (m/s)		
	Coastal	**Inland**	**Difference**	**Coastal**	**Inland**	**Difference**	**Coastal**	**Inland**	**Difference**
Tokyo	17.6	16.0	1.6	32.2	35.8	−3.6	5.1	2.3	2.8
Osaka	17.9	16.6	1.3	32.5	34.2	−1.7	3.0	1.9	1.1
Nagoya	17.5	16.8	0.7	32.0	33.5	−1.5	3.5	2.2	1.3
	Enthalpy (J)			**SET*(°C)**			**WBGT (°C)**		
	Coastal	**Inland**	**Difference**	**Coastal**	**Inland**	**Difference**	**Coastal**	**Inland**	**Difference**
Tokyo	77,413	77,148	265	33.6	35.2	−1.6	31.0	31.4	−0.4
Osaka	78,423	76,949	1474	33.9	34.4	−0.5	31.2	31.1	0.1
Nagoya	76,859	76,795	64	33.4	34.0	−0.6	30.9	31.1	−0.2
	PET (°C)								
	Coastal	**Inland**	**Difference**						
Tokyo	35.4	38.4	−3.0						
Osaka	36.1	37.5	−1.4						
Nagoya	35.6	37.1	−1.5						

PET, *physiological effective temperature;* SET*, *new Standard effective temperature;* WBGT, *Wet-bulb globe temperature.*

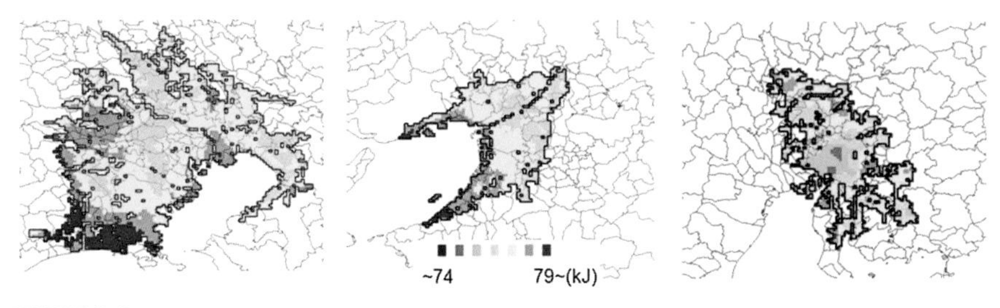

FIGURE 3.5

Enthalpy distribution at 2 m high in Tokyo, Osaka, and Nagoya at 14:00 averaged on fine summer sea breeze day.

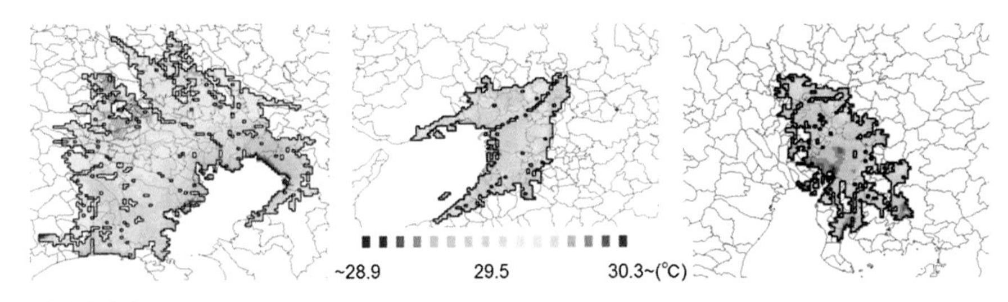

FIGURE 3.6

Wet-bulb globe temperature (WBGT) distribution at 2 m high in Tokyo, Osaka, and Nagoya at 14:00 averaged on fine summer sea breeze day.

due to a slightly high absolute humidity. However, the difference is within 2% of the total amount, so the influence on cooling energy consumption is sufficiently small. In Fig. 3.6, WBGT is nearly the same at coastal area with low air temperature and high absolute humidity and at inland area with high air temperature and low absolute humidity. It is slightly higher in inland area and confirmed especially in Tokyo which has wide inland area. Fig. 3.7 shows that SET* near the coast is slightly lower than that in the inland due to high wind velocity, whereas, in Fig. 3.8, PET near the coast is slightly lower than that in the inland due to low air temperature and high wind velocity. Humidity sensitivity to PET is small. Mean radiant temperature, which is a great influence on WBGT, SET*, and PET, changes greatly due to the influence

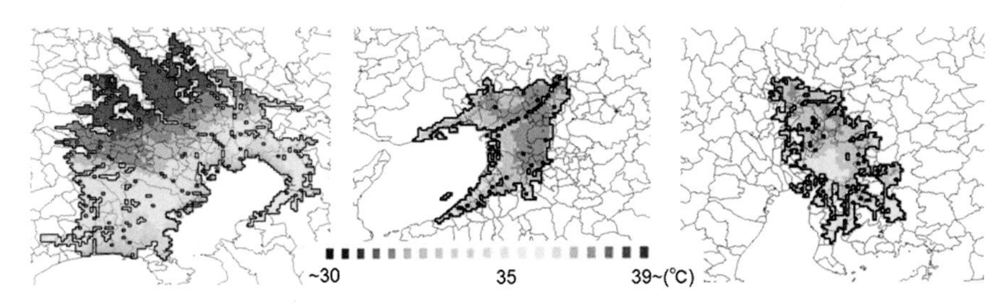

FIGURE 3.7

New Standard effective temperature (SET*) distribution at 2 m high in Tokyo, Osaka, and Nagoya at 14:00 averaged on fine summer sea breeze day.

FIGURE 3.8

Physiological effective temperature (PET) distribution at 2 m high in Tokyo, Osaka, and Nagoya at 14:00 averaged on fine summer sea breeze day.

of solar radiation, but there is no difference between coastal and inland area. As mentioned above, the difference of enthalpy and WBGT between near the coast and inland is small, because the influence by air temperature and absolute humidity is canceled. In the coastal area where the sea breeze enters, as the wind velocity of upper air is large, the wind velocity inside the street canyon also tends to be large. As described below, while the wind velocity inside the street canyon varies influenced by urban block form, it was assumed as 0.2 of upper wind velocity for Figs. 3.7 and 3.8.

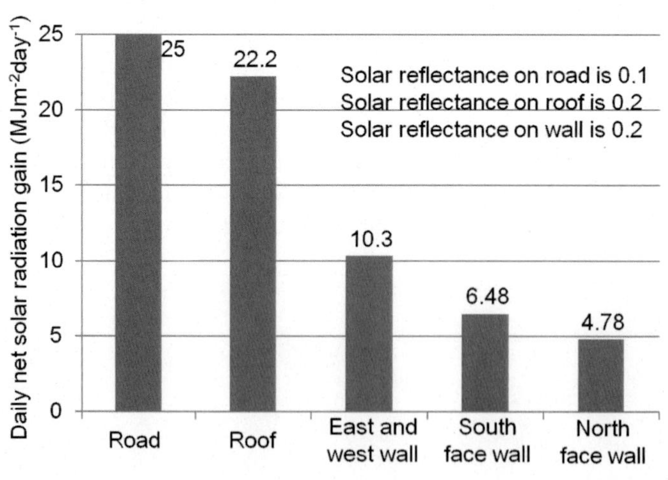

FIGURE 3.9

Daily net solar radiation gains in summer on each surface of an isolated building.

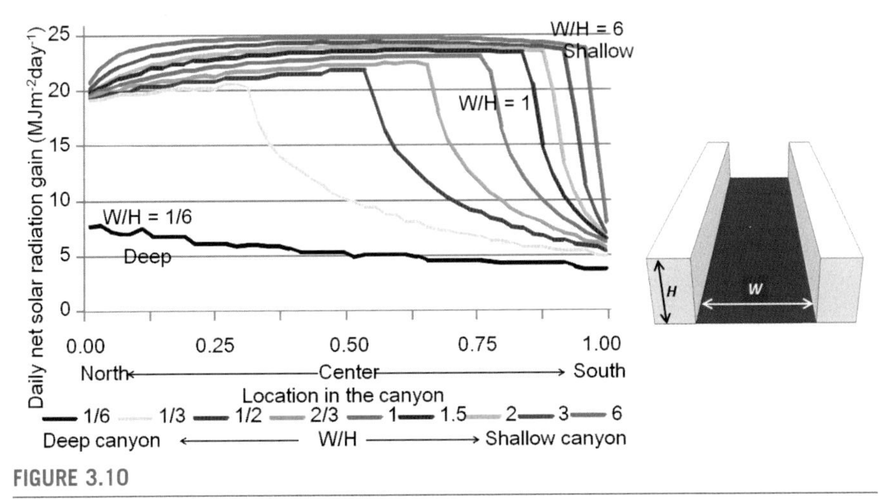

FIGURE 3.10

The daily net solar radiation gain for various aspect ratios on an east–west road.

3. Urban morphology and radiation environment in the street canyon

3.1 Features of radiation environment [5]

The place that has received much solar radiation is likely to be a "hot spot." Daily net solar radiation gains in summer on each surface of an isolated building are shown in Fig. 3.9. Lower solar reflectance is the reason that this gain on the road is greater than that on the roof. Daily net solar radiation on the road and roof is higher than that on the wall because of the greater solar altitude. Daily net solar radiation on the east and west surfaces is larger than that on the north and south surfaces of

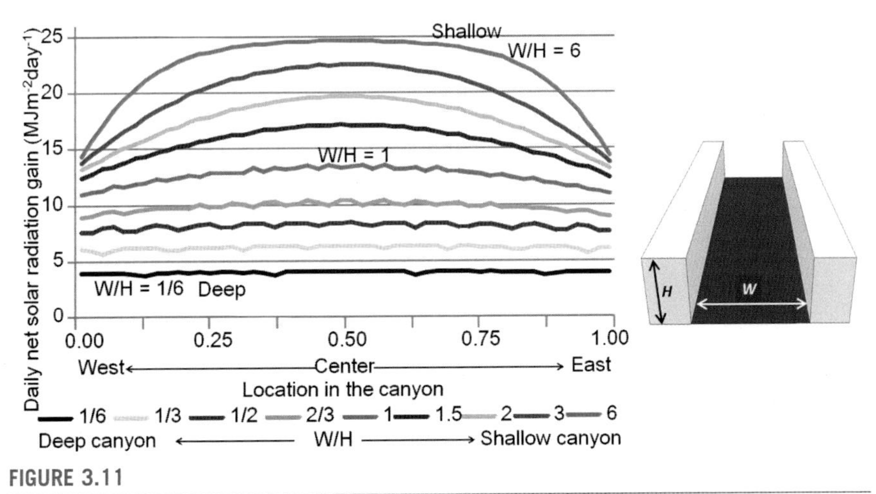

FIGURE 3.11

Daily net solar radiation gains for various aspect ratios on a north–south road.

the wall. Therefore, the roof and road surfaces have a higher priority as heat island mitigation locations under general summer conditions.

The daily net solar radiation gain on an east–west road is shown in Fig. 3.10. Road width W and building height H are varied as a reference aspect ratio W/H of 1 ($W = H = 15$ m). The distributions are dominated by the shadows of the buildings to the south. High-priority areas for urban heat island mitigation measures extend to the buildings on the south sides of the roads for street canyons with larger W/H ratios.

Daily net solar radiation gains on north–south road are shown in Fig. 3.11. The distribution is dominated by the shadows of the buildings on both sides of the road. Other priority locations for urban heat island mitigation are the areas around the centers of roads for models with larger W/H ratios.

Daily net solar radiation gains at intersections are an integrated distribution of those shown in Figs. 3.10 and 3.11. The amounts of daily net solar radiation gain in the areas close to the north sides are large, as shown in Fig. 3.10. Furthermore, those in the southern areas are large around the centers of the roads owing to the influence of the shadows of southwest and southeast buildings, as shown in Fig. 3.11.

Daily net solar radiation gains on road affected by taller adjacent building are shown in Figs. 3.12 and 3.13. It is assumed that a 15, 30, or 60 m taller building is located nearby the objective road, and the horizontal axis indicates the distance from the taller building in the north–south direction in Fig. 3.12 and the west–east direction in Fig. 3.13. The daily net solar radiation gain on the road located on the north side of a taller building is greatly reduced in the vicinity of the taller building, but the affected area is relatively small. The radiation gain on the road on the west or east side of a taller building is generally reduced at short distances from the taller building, and the affected area is somewhat larger. The low solar altitudes in the morning and

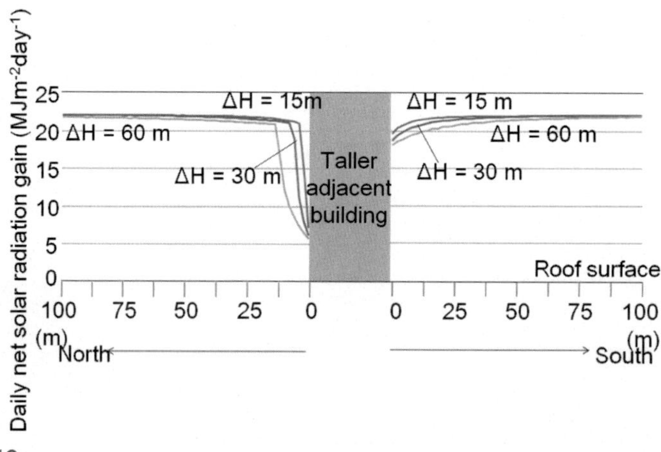

FIGURE 3.12

Effects of a taller building adjacent to the north or south side of the objective building on solar radiation gains.

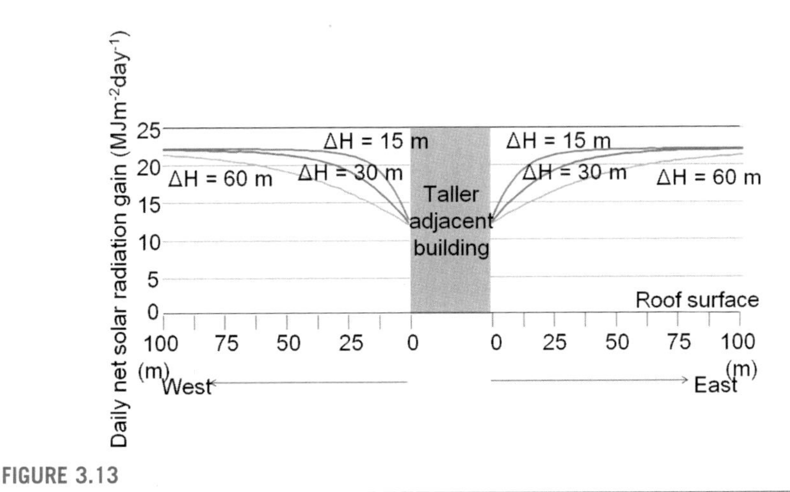

FIGURE 3.13

Effects of a taller building adjacent to the west or east side of the objective building on solar radiation gains.

evening periods affect the daily solar gains on adjacent shaded roads. Even if a 60 m taller building is located on the south side of the target road, the shadow effect caused by the taller building is small when the distance from the taller building is more than 15 m. If the objective road is located on the west or east side of the taller building, the affected area varies depending on the taller building height. In addition, if the objective road is located to the northwest or northeast of a taller building, the affected area is formed by combining Figs. 3.12 and 3.13.

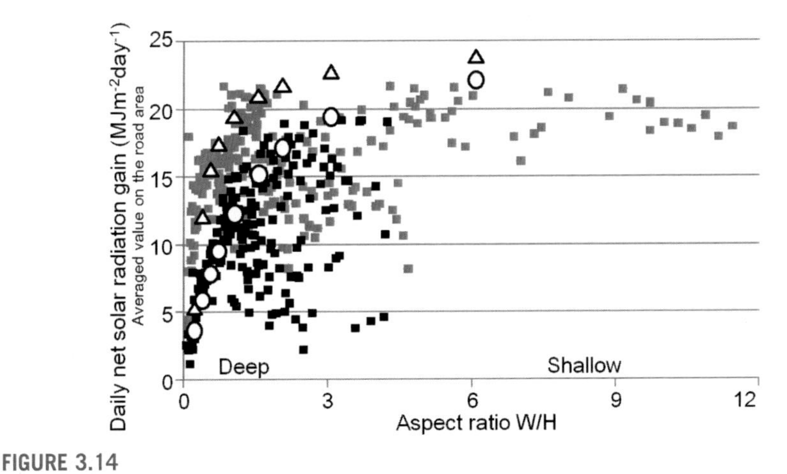

FIGURE 3.14

Aspect ratios and daily net solar radiation gains on actual and modeled roads. ■: Actual E–W road, ■: Actual N–S road, Δ: Simple model E–W road, O: Simple.

Relationships between the aspect ratio and the daily net solar radiation gain on actual and modeled roads are shown in Fig. 3.14. The daily net solar radiation gains are averaged values for each road. The building height H is the building's height along the road. The results of the simple urban canyon model with uniform building heights are also shown in Fig. 3.14. The daily net solar radiation gains are large for roads where W/H is more than about 1.5 and there is less variation among the north—south roads than among the east—west roads. As they are affected by the shade of surrounding buildings, for north—south roads in particular, the amounts of daily net solar radiation gain are extremely small. The effects of high-rise buildings are not represented in the simple urban canyon model with uniform building heights. Adjacent high-rise buildings substantially affect neighboring buildings, and high-rise buildings also affect their neighbors to the northeast and northwest. Accordingly, when considering the locations for urban heat island mitigation measures, we should take into account not only the daily net solar radiation gains represented by the aspect ratios as shown in Figs. 3.10 and 3.11, but also the influence of high-rise buildings as shown in Figs. 3.12 and 3.13.

3.2 Appropriate selection of heat island measure technologies [6].

The outline of the street canyon model with a street tree and green wall is shown in Fig. 3.15. Both side buildings with 30 m height were located 1 m away from the road, and the sidewalk with 3-m width was in front of them. A street tree with 2 m width was located between the roadway and sidewalk. The widths of the roadway were 10, 20, and 40 m. Objective heat island measure technologies were a street tree, a green wall, high-reflectance paint, and water-retentive pavement. The height of a street tree, the height under the tree crown, solar radiation shielding factor of leaves, and evaporative efficiency of leaves were 10, 2.5, 0.8, and 0.3 m, respectively.

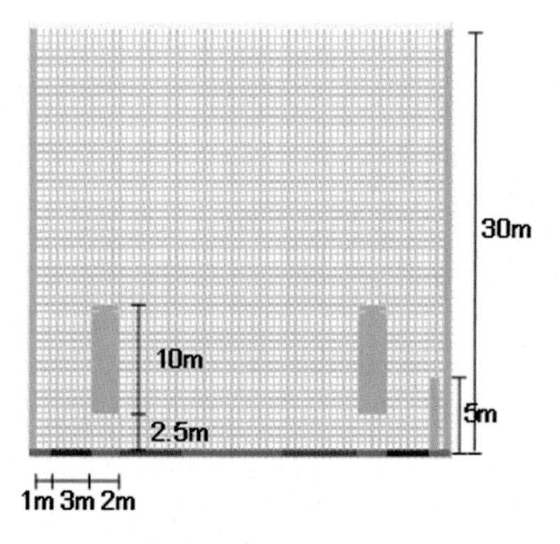

FIGURE 3.15

Outline of the street canyon model with a street tree and green wall.

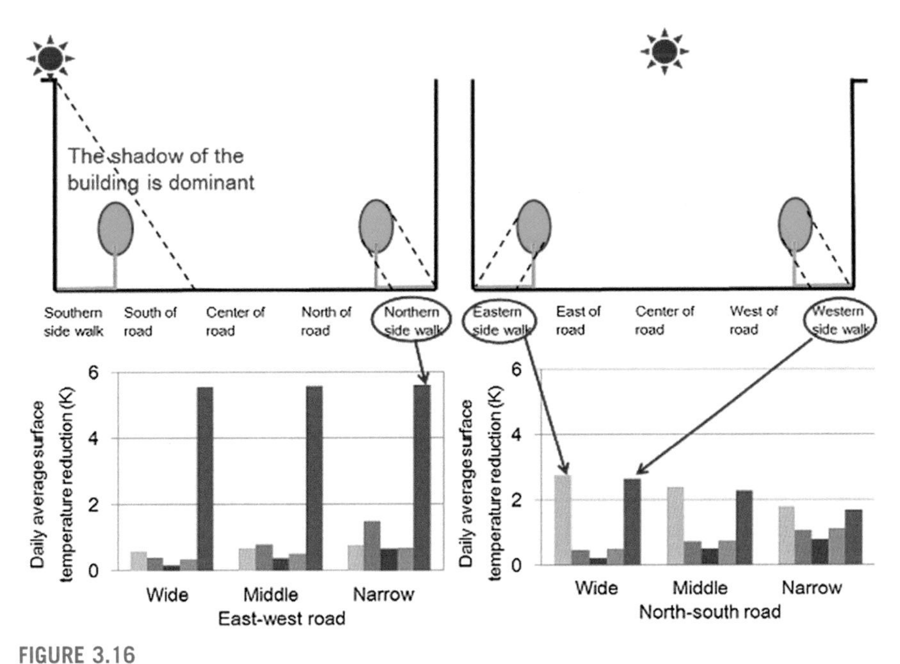

FIGURE 3.16

Daily average surface temperature reduction due to a street tree (left: east–west road, right: north–south road).

The green wall was 5 m high and 0.5 m away from the concrete wall. Solar radiation shielding factor and evaporative efficiency of leaves were 0.8 and 0.3, respectively. The reflectance of high-reflectance paint was 0.4, and it was painted on the asphalt road surface. The evaporative efficiency of the water-retentive pavement was 0.3, which was different from that of the asphalt road surface. The pavement was assumed to be of a continuous water supply type with irrigation equipment.

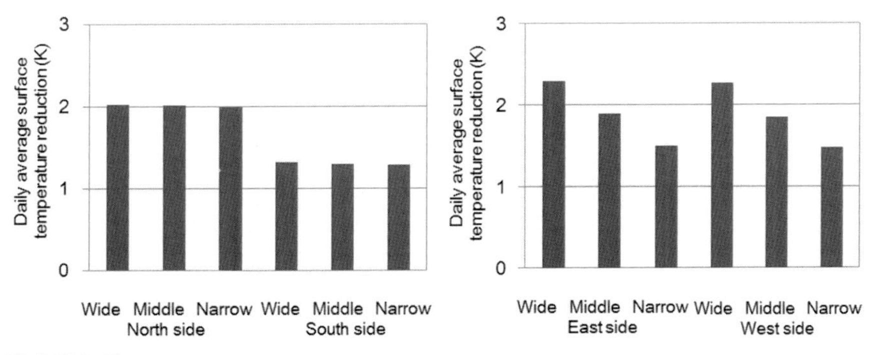

FIGURE 3.17

Daily average surface temperature reduction due to a green wall (left: east–west road, right: north–south road).

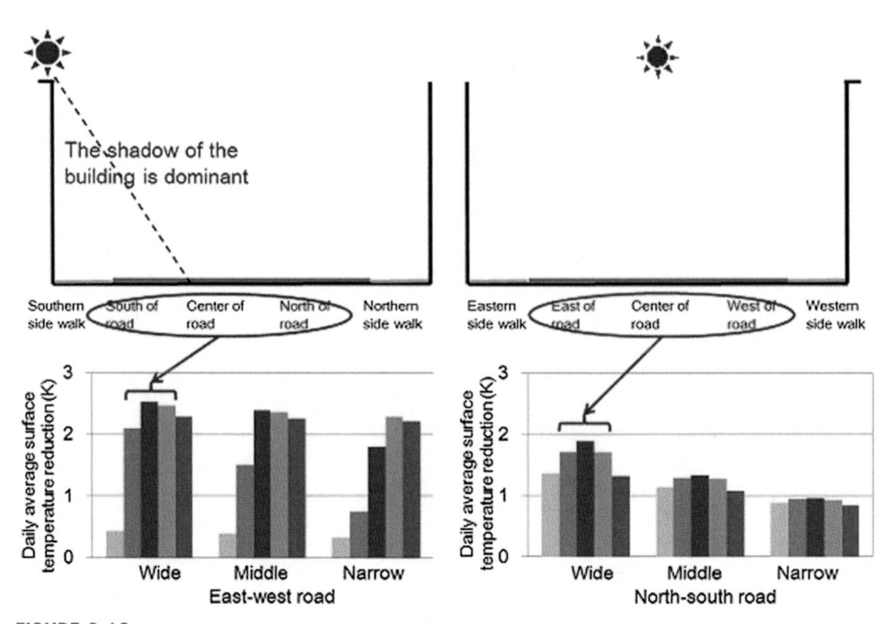

FIGURE 3.18

Daily average surface temperature reduction due to high-reflectance paint (left: east—west road, right: north—south road).

The daily average surface temperature reduction caused by the street tree is shown in Fig. 3.16. The benefits on the northern sidewalk of the east—west road and the eastern and western sidewalk of the north—south road were observed to be the greatest. This meant that the improvement of the thermal environment of the pedestrian space was remarkable.

The daily average surface temperature reduction caused by the green wall is shown in Fig. 3.17. The benefit of the east—west road due to the north-side wall

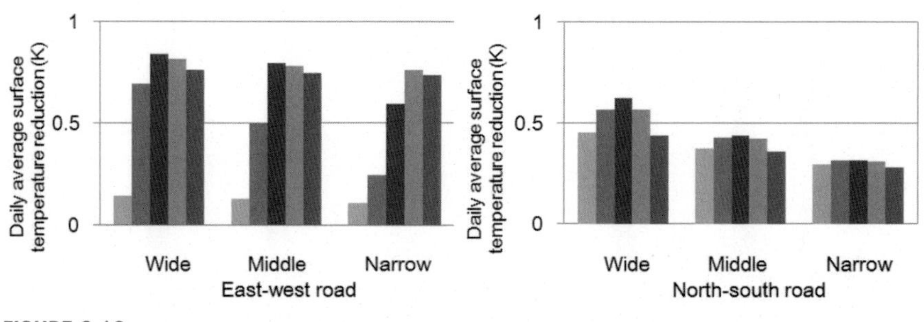

FIGURE 3.19

Daily average surface temperature reduction due to water-retentive pavement (left: east—west road, right: north—south road).

was larger than that due to the south-side wall. However, the difference due to the road width was uncertain. The benefit difference between the west- and east-side walls for the north–south road was small. Moreover, larger road width yielded larger benefit. The wall that received greater solar radiation also returned a larger benefit.

Daily average surface temperature reduction due to high-reflectance paint and water-retentive pavement are shown in Figs. 3.18 and 3.19. The benefits due to both high-reflectance paint and water-retentive pavement are larger at the center of the north–south road and from the center to the northern side of the east–west road, where the solar radiation gain is large. The daily average surface temperature reduction due to a combination of the street tree, high-reflectance paint, and water-retentive pavement is shown in Fig. 3.20. As previously described, the benefit due to both high-reflectance paint and water-retentive pavement appears mainly on the

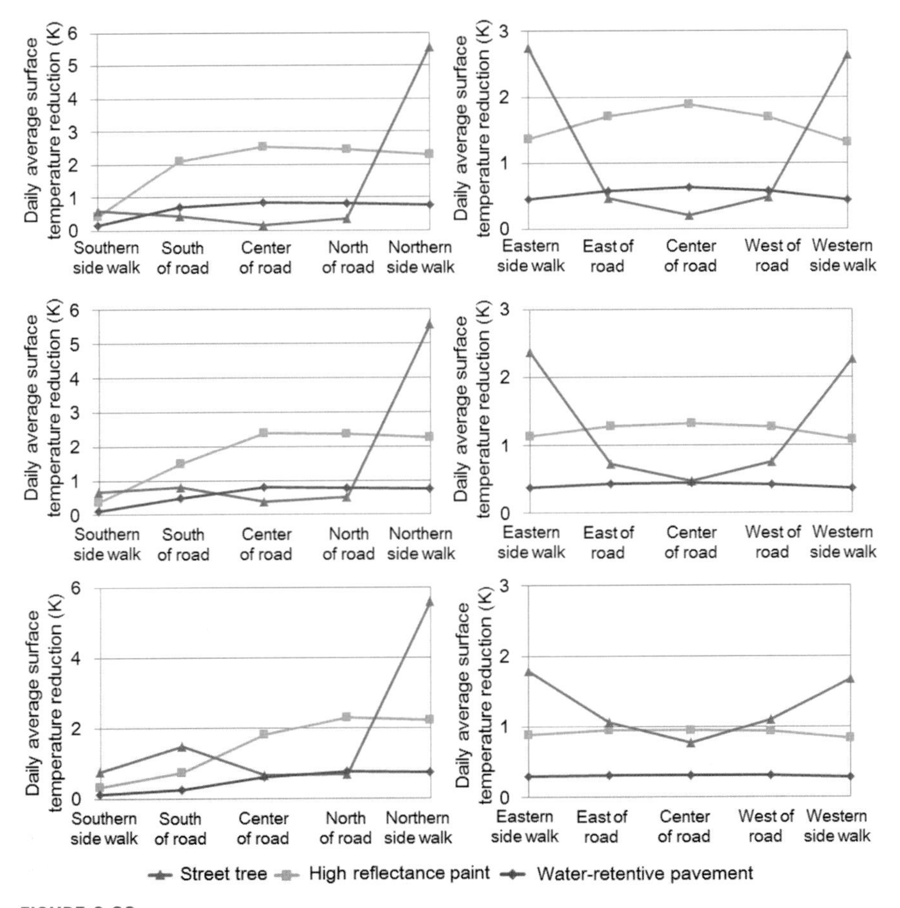

FIGURE 3.20

Daily average surface temperature reduction due to a street tree, high-reflectance paint, and water-retentive pavement upper-left: wide east–west road, upper-right: wide north–south road, medium-left: middle east–west road, medium-right: middle north–south road, lower-left: narrow east–west road, lower-right: narrow north–south road).

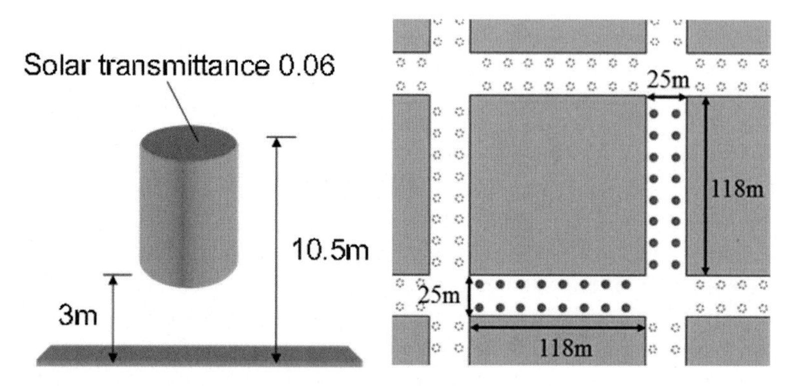

FIGURE 3.21

Street tree model and objective urban block (building height is 18 m).

roadway. On the other hand, the benefit due to street tree is concentrated mainly on the sidewalk. This is because the benefit from the street tree is caused by solar radiation shielding by the tree crown, and their shadows occur mainly on the sidewalk. Therefore, it can be inferred that the introduction of water-retentive pavement and high-reflectance paint on the roadway is more appropriate for the road surface temperature reduction, while the introduction of a roadside tree is more appropriate for improving the thermal environment of the sidewalk. Technology such as wall-based solar radiation shielding (e.g., green wall) is suitable for the temperature reduction of a wall surface.

The relationship between the street tree layout and the thermal environmental improvement was analyzed for the appropriate introduction policy of roadside trees. The calculation results of mean radiant temperature (MRT) at 12:00, 15:00, and 17:00 on a typical sunny day in summer were used for the evaluation. The shape and solar transmittance of the tree crown were calculated from the survey results of street trees in Kobe city. The street tree model and objective urban block are shown in Fig. 3.21. It was assumed that street crown is floating in air; the solar transmittance is uniform for the entire crown and does not change with respect to time. The building height was set to 18 m. Direct and diffuse solar radiation was calculated, but reflected solar radiation was not taken into account. Street tree layout parameters are listed in Table 3.3, and the location of street trees is shown in Fig. 3.22. The diameter of the cylindrical tree crown in the street tree model is the tree crown width (A) and the distance between two adjacent cylinders is the tree interval (B).

Table 3.3 Parameters of the street tree layout.

Tree crown width (A)	Tree interval (B)
4 m	6, 8, 10, 12 m
6 m	8, 10, 12 m

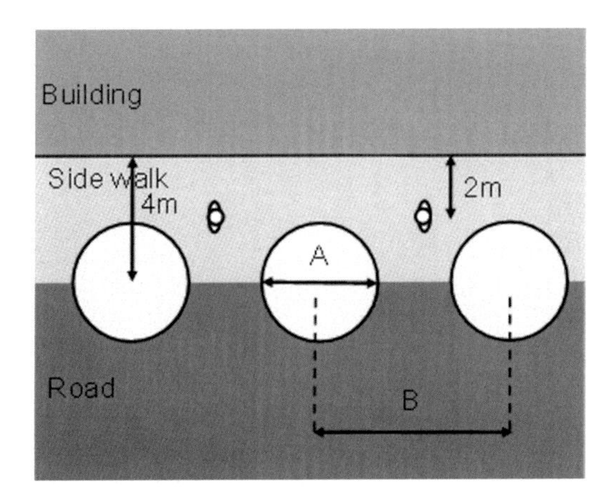

FIGURE 3.22

Location of street trees (A: crown width, B: tree interval).

The solar radiation gain averaged along the street at 12:00 in several tree interval cases is shown in Fig. 3.23. For large tree crown width and small tree intervals, solar radiation gain was small owing to the large proportion of the shaded area. The reduction in solar radiation gain on the northern sidewalk of the east–west road was greater than that on the eastern sidewalk of the north–south road because half of the roadside-tree shadow falls on the driveway of the eastern sidewalk of the north–south road.

The solar radiation gain averaged along the street at 15:00 in several tree interval cases is shown in Fig. 3.24. As the solar radiation was less than that at 12:00, the reduction in solar radiation gain due to the change of street tree layout was also small, but the trend of the reduction of solar radiation gain was the same as that observed at 12:00. The reduction of solar radiation gain on the northern sidewalk of the east–west road was greater than that of the eastern sidewalk on the north–south road, at 15:00 as

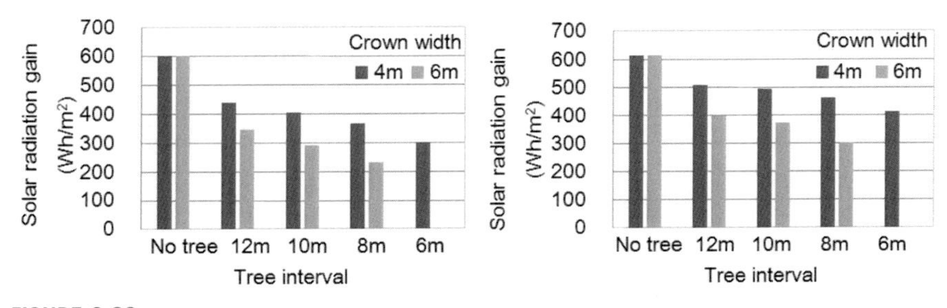

FIGURE 3.23

Solar radiation gain averaged along the street at 12:00 for several tree interval cases (left: northern sidewalk of east–west road, right: eastern sidewalk of north–south road).

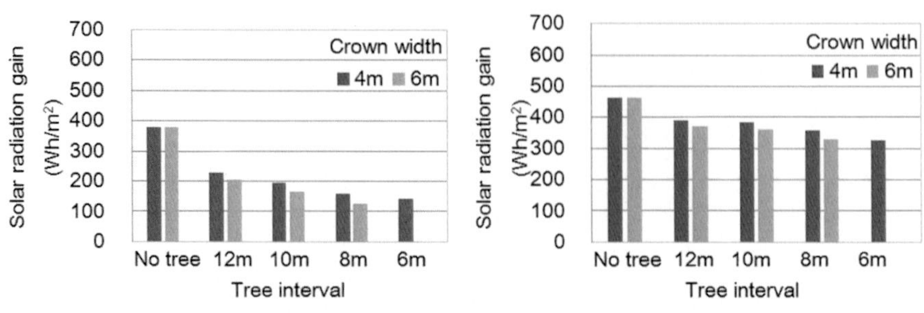

FIGURE 3.24

Solar radiation gain averaged along the street at 15:00 for several tree interval cases (left: northern sidewalk of east–west road, right: eastern sidewalk of north–south road).

well as 12:00, because much of the shadow falls on the northern sidewalk of the east–west road rather than on the eastern sidewalk of the north–south road.

The reduction of the solar radiation gain on the sidewalk depends on the area and the position of the shadow of the street trees. The shadow tends to fall on the northern sidewalk of the east–west road rather than the eastern sidewalk of the north–south road because the shadow on the eastern side of the north–south road falls on the wall and roadway as opposed to the northern side of the east–west road.

MRT averaged along the street at 15:00 for several tree interval cases is shown in Fig. 3.25. The tree crown width was set to 4 m. The calculation results considering only the influence of long-wave radiation and those including short-wave radiation are separately shown. The human body was assumed to be a cube, the weighting

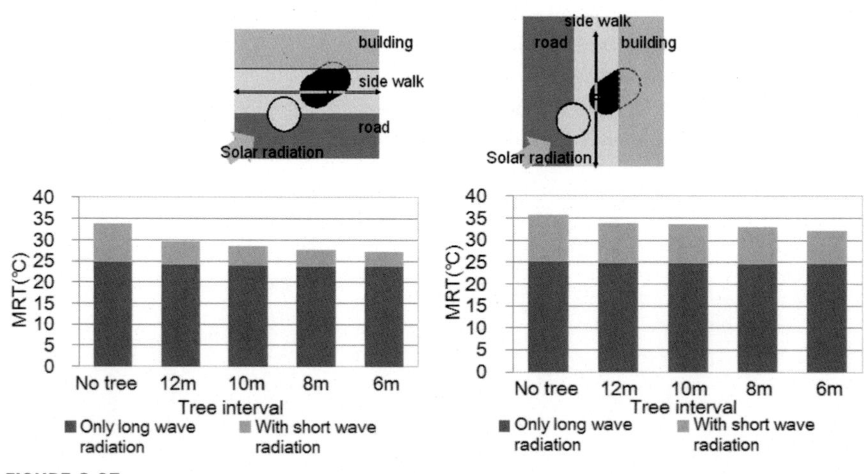

FIGURE 3.25

Mean radiant temperature (MRT) averaged along the street at 15:00 for several tree interval cases (crown width is 4 m) (left: northern sidewalk of east–west road, right: eastern sidewalk of north–south road).

factor for the upper and lower surfaces was 0.024, and the weighing factor for the side surface was 0.238. For solar radiation gain area, the solar absorbance of the human body was 0.5, where the human body walks on the center of the sidewalk and MRT is averaged along the sidewalk.

The influence of the reduction in short-wave radiation gain was significant for MRT. The influence of the difference in street tree layout was mainly due to the difference in short-wave radiation falling on the human body. The reduction in MRT was greater on the northern sidewalk of the east–west road than on the eastern sidewalk of the north–south road. Because the shadows of the roadside trees fall on the direction perpendicular to the eastern sidewalk of the north–south road and on the direction parallel to the northern sidewalk of the east–west road, the area where the human body is under the influence of a shadow is large on the northern sidewalk of the east–west road.

The reduction in MRT on the sidewalk depends on the area where the human body is under the influence of a shadow. The shadow tends to fall on the direction perpendicular to the eastern sidewalk of the north–south road and on the direction parallel to the northern sidewalk of the east–west road. Therefore, the shadow area on the northern sidewalk of the east–west road was larger than on the eastern (western) sidewalk of the north–south road.

As the solar radiation incident on a building façade is small compared with that on roof and pavement, the priority for cooling measures on building façades is low. In addition, high-reflectance technology applied to building façades reflects solar radiation downward as well as upward. Fig. 3.26 shows reflection from a building façade [7]. A simple, two-dimensional building façade was assumed. The incident angle of the direct solar radiation is specified by the sun angle (a function of date, time, and latitude), regardless of the orientation of the objective façade. The diffuse reflection of solar radiation is taken as half upward and half downward from the horizontal. Reflectance values of 0.25 for conventional materials and 0.75 for high-reflectance materials were assumed. As the diffuse reflection does not depend on the sun angle, the change of the sun angle may not be taken into account. Additionally, as the most significant impact on pedestrians near the wall by the specular

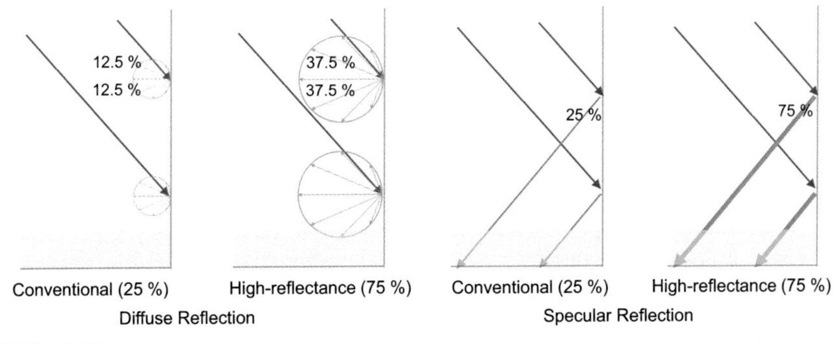

FIGURE 3.26

Diffuse and specular reflection from a building façade.

reflection occurs in the high sun angle condition, the influence by the specular reflection should be considered in the high sun angle condition.

In the case of a completely diffuse surface, such as concrete, paint, or wood, used as a conventional building façade, 12.5% of the solar radiation is reflected upward and 12.5% is reflected to a pedestrian area. Approximately 75% of solar radiation is absorbed by the building façade and is redistributed as sensible or radiative heat flux into the street canyon and to the building interior, which increases the indoor cooling load. In total, 87.5% of solar radiation is converted into heat near the ground surface.

In the case of a high-reflectance but completelydiffuse building façade, 37.5% of solar radiation is reflected upward and 37.5% is reflected to a pedestrian area. Approximately 25% of the solar radiation is absorbed by the building façade and is redistributed as sensible or radiative heat flux into the street canyon and to the building interior. In total, 62.5% of solar radiation is converted into heat near the ground surface. Therefore, heat generation near the ground surface is reduced 25% when the solar reflectance of a completelydiffuse building façade changes from 0.25 to 0.75.

In the case of a completelyspecular surface, such as metal, glass, or tile, used as a conventional building façade, 0% of solar radiation is reflected upward and 25% is reflected to a pedestrian area. 75% of solar radiation is absorbed by the building façade and is redistributed as sensible or radiative heat flux into the street canyon and to the building interior. In total, 100% of solar radiation is converted into heat near the ground surface.

In the case of a completely specular high-reflectance building façade, 0% of solar radiation is reflected upward and 75% of solar radiation is reflected to a pedestrian area. The rest, i.e., 25% of solar radiation is absorbed by the building façade and is redistributed as sensible or radiative heat flux into the street canyon and to the building interior. In total, 100% of solar radiation is converted into heat near the ground surface. As specular materials introduce such large amounts of heat near the ground surface, cool façades are recommended to reduce heat island effects.

The solar radiation reflected to a pedestrian area is increased 25% in the diffuse surface case and 50% in the specular surface case when the solar reflectance of the building façade changes from 0.25 to 0.75. The impact of solar radiation reflected from the building façade to a pedestrian must, therefore, be further evaluated. The height at which the specular reflection of solar radiation from the building façade does not affect a pedestrian (A = 2 m, B = 1.5 m) at 12:00 on August 1 in Osaka was calculated as 8.3 m from the ground, or above the fourth floor if the height of each floor is 3.5 m. As specular reflection is likely to occur at high sun angles, it is recommended that high-reflectance technology be used on building façades above the fourth floor in this particular case (Fig. 3.27).

Adverse effects on a pedestrian from the reflection of solar radiation are an issue when high-reflectance technology is applied to building façades. Retroreflective materials have been studied as a way to address this problem. The shape and reflection characteristics of retroreflective tile and the effects of using retroreflective material on the lower floors are shown in Fig. 3.28. A retroreflective tile [8] has been already

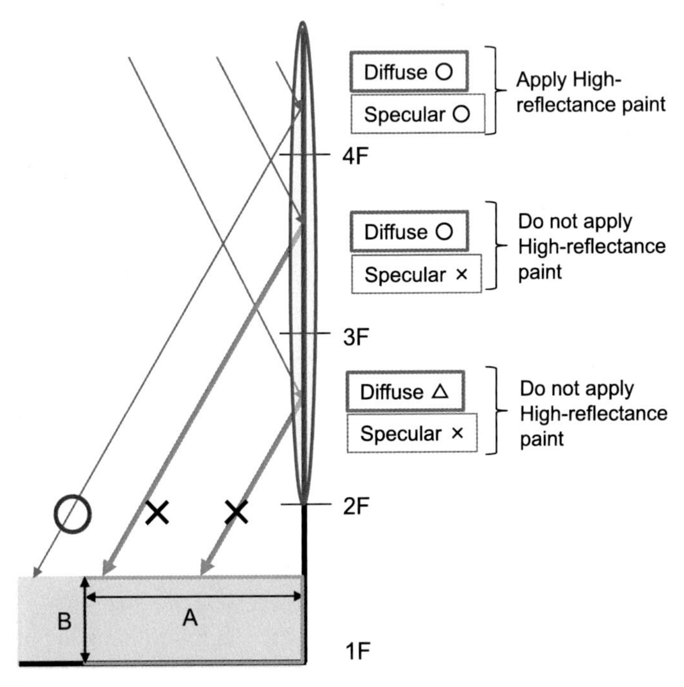

FIGURE 3.27

The height at which specular reflection of solar radiation from the building façade does not affect pedestrians (A = 2 m, B = 1.5 m, at 12:00 on August 1 in Osaka).

commercially available in Japan. Only a very small fraction of solar radiation is reflected downward from the retroreflective material.

Pedestrians are not affected by retroreflective material that is applied to the building façade below the third floor as long as the solar radiation is reflected completely

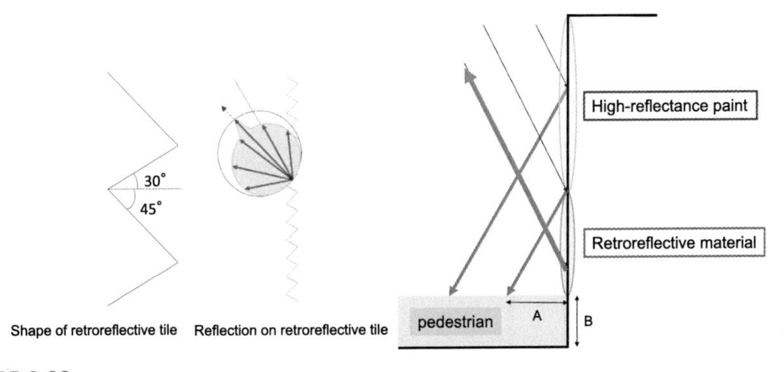

FIGURE 3.28

Shape and reflection characteristics of retroreflective tile (left) and the effects of using retroreflective material on the lower floor (right).

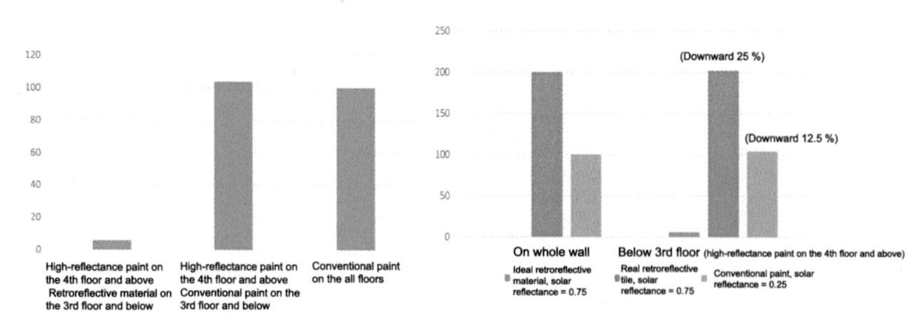

FIGURE 3.29

The ratio of solar radiation reflected to a pedestrian for an idealized retroreflective material (left) and a realistic retroreflective material (right).

upward. Only the diffuse reflection from high-reflectance technology applied to the building façade above the fourth floor would affect pedestrians. The amount of solar radiation reflected to a pedestrian from an ideal retroreflective material is calculated as a ratio to that of a conventional building façade with solar reflectance of 0.25 (Fig. 3.29). A solar reflectance of 0.75 is used for both high-reflectance materials and retroreflective materials. Results were calculated for two cases. In the first case, high-reflectance technology was applied to the fourth floor and above, while a conventional façade was used on the third floor and below. The calculated value for this case was 104. In the second case, high-reflectance technology was applied to the fourth floor and above while retroreflective material was applied to the third floor and below. The calculated value for this case was 6.

The retroreflective tile used for this investigation consists of an upward slope of 45 degrees to horizontal and a downward slope of 30 degrees to horizontal (Fig. 3.28). As a part of solar radiation is reflected diffusely from each slope surface, it is not a completely retroreflective material. Even so, a relatively large amount of solar radiation is reflected upward. When incident solar radiation is 60 degrees to the objective building façade, almost two-thirds of solar radiation is reflected upward and another one-third is reflected downward. Therefore, 50% of solar radiation is

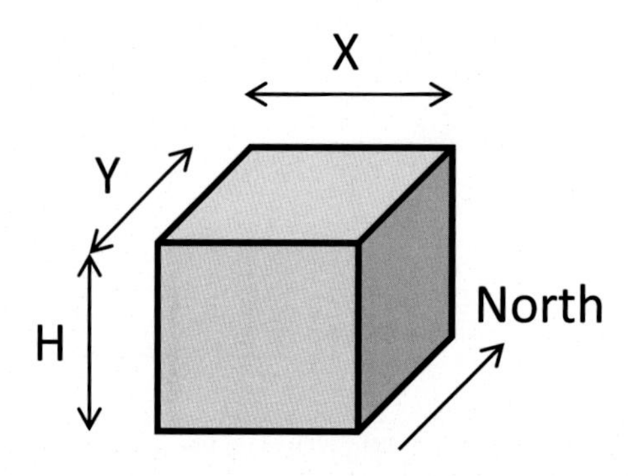

FIGURE 3.30

Simple building model with height H, east–west width X, and north–south width Y.

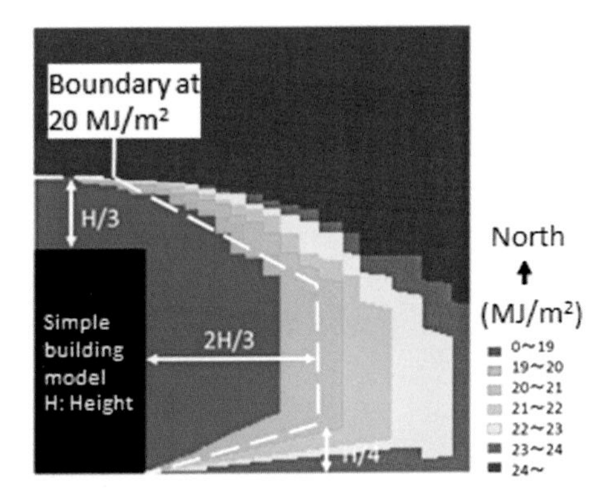

FIGURE 3.31

Distribution of daily integrated solar radiation around the simple building model (the white dashed line is marking the boundary at 20 MJ/m2, which correspond to the 80% of the maximum daily integrated solar radiation).

reflected upward and 25% of solar radiation is reflected downward from an overall solar reflectance of 0.75 for retroreflective tile. For comparison, when the overall solar reflection of high-reflectance paint is 0.75, 37.5% of solar radiation is reflected both upward and downward. When the overall solar reflectance of a conventional façade is 0.25, 12.5% of solar radiation is reflected both upward and downward. Therefore, the effect on pedestrians from using retroreflective materials is similar to the effect from using high-reflectance paint.

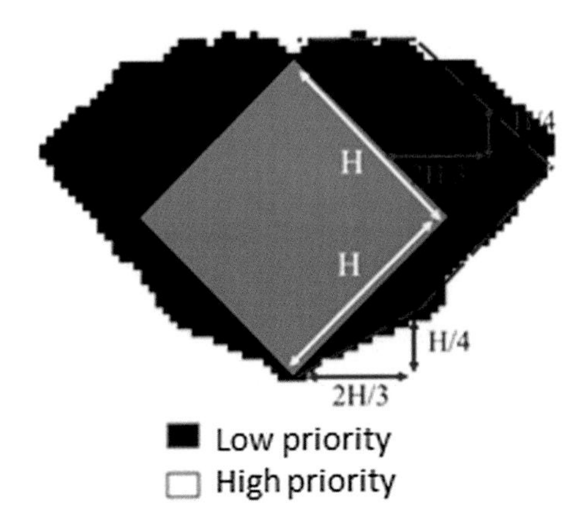

FIGURE 3.32

Priority of adopting adaptation measures around a simple building model with an orientation of 45degrees with respect to the west direction.

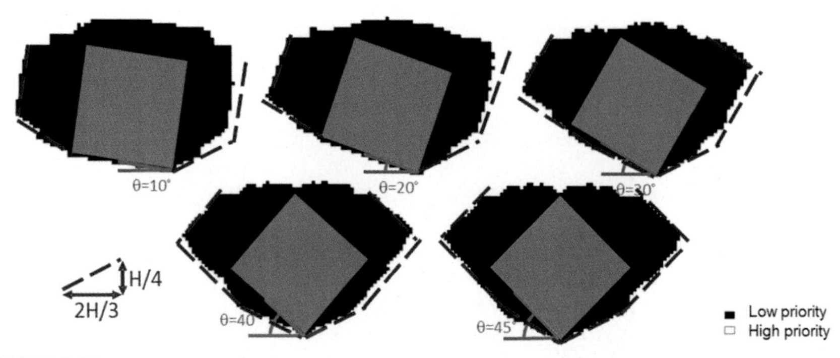

FIGURE 3.33

Priority of adopting adaptation measures around a simple building model. The building orientation presents an increasing angle, θ, with respect to the west direction.

3.3 Selection criteria of hot spot [9]

A simple building model with height *H*, east to west width *X*, and north to south width *Y* is shown in Fig. 3.30. Distribution of the daily integrated solar radiation around a simple building model is shown in Fig. 3.31. It is calculated based on the orbit of the sun for a sunny, summer day in Osaka city. The boundary (white dashed line) marking the 20 MJ/(m^2day) (which corresponds to the 80% of the maximum daily integrated solar radiation) is located at about *H*/3 from the southern building wall and at about 2*H*/3 from the eastern and western walls with a gradient of about *H*/4 from the southern tip of the building.

The priority of adopting adaptation measures around a simple building model oriented at 45 degrees and various angles with respect to the west direction are shown in Figs. 3.32 and 3.33. The priority is low in areas placed at a distance of 2*H*/3 from any of the eastern and western walls of the building, with a gradient of about *H*/4 toward the northern side. As shown in Fig. 3.33, the low-priority area in the northern side of the building decreases in size as the angle θ between the building and the west direction approaches 45 degrees.

Table 3.4 Calculation condition for CFD model.

Turbulence model	Standard k-ε model
Advection term	Up-wind difference scheme
Inflow boundary	Power law, 3.2 m/s at 54 m high, power: 0.25
Outflow boundary	Zero gradient condition
Up, side boundary	Free-slip condition
Wall, ground boundary	Generalized log-law
Grid resolution	10 m (x), 10 m (y), 1 m (z) in the target area

CFD, *Computational fluid dynamics.*

FIGURE 3.34

Calculation results of wind velocity at the height of 2 m.

4. Urban morphology and wind environment in the street canyon

4.1 Features of wind environment [10].

The standard k-ε turbulence model (one of the RANS models) was selected for use in the simulation. General-purpose computational fluid dynamics (CFD) software (STREAM, version 8, Software Cradle Co., Ltd.) was used for calculation. The Navier–Stokes equations were discretized using a finite volume method, and the SIMPLE algorithm was used to handle pressure–velocity coupling. The calculation conditions are shown in Table 3.4, referring to Tominaga et al. [11]. The applicability of this CFD software for an urban area such as Osaka City has been verified using a verification database provided by Tominaga et al. [11].

The calculated results of wind velocity 2 m above the ground are shown in Fig. 3.34. Based on measurement results at Osaka observatory, wind speed (3.2 m/s) for westerly wind 54 m above the ground, with a power law vertical profile (power: 0.25), were the inflow boundary conditions set. Grid resolution is 10 m, 10 m, 1 m. Data about individual building shapes handled using a GIS tool were provided by Osaka City office. The height of each building was calculated by multiplying the floor height from each building use by the number of stories specified in the data of each building. Smaller objects (e.g., trees, signs, cars, human bodies) could not be reproduced. Building shape information of the adjacent city was added based on map data and aerial photographs. Wind

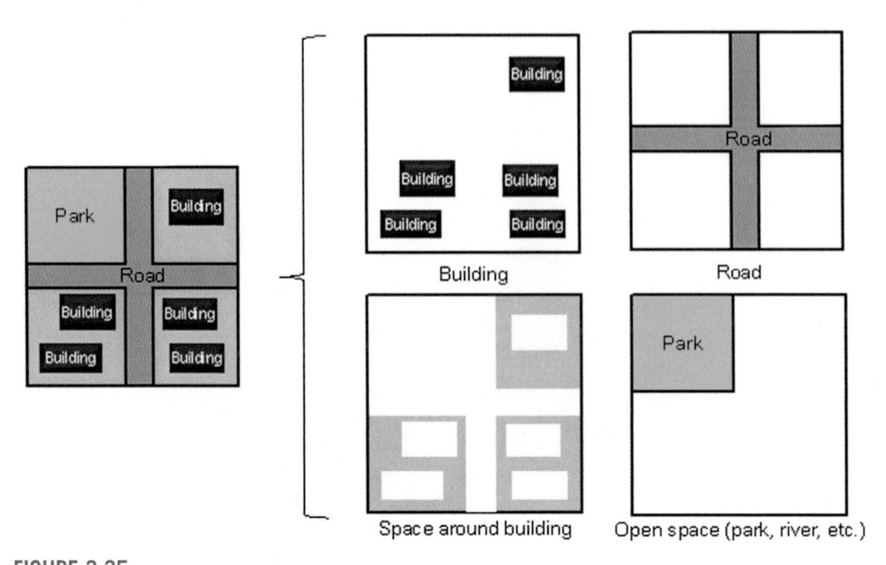

FIGURE 3.35

Classification of urban block components.

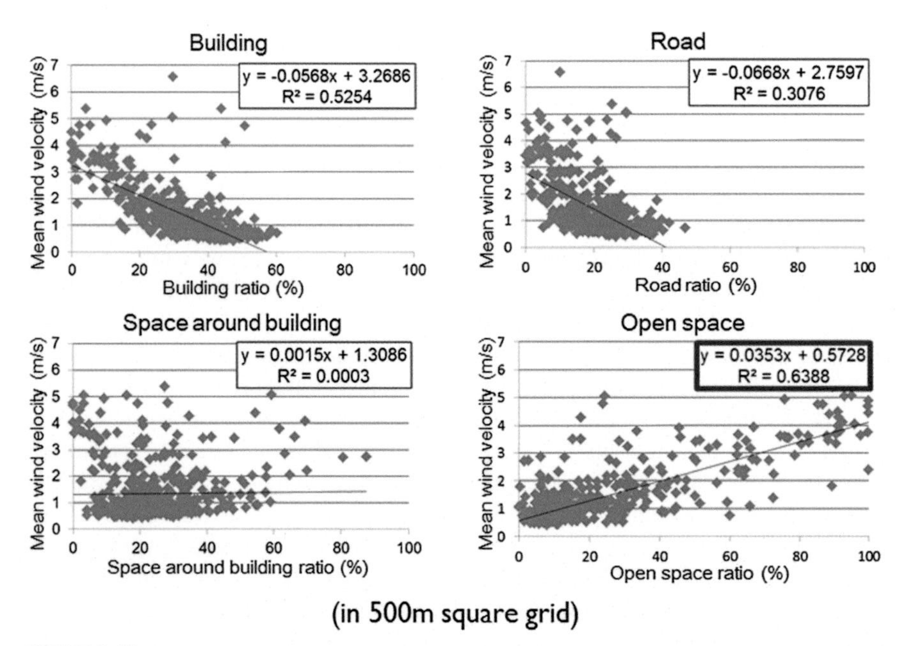

FIGURE 3.36

Relationship between urban block component ratio and mean wind velocity at the height of 2 m, averaged over grids 500 m square.

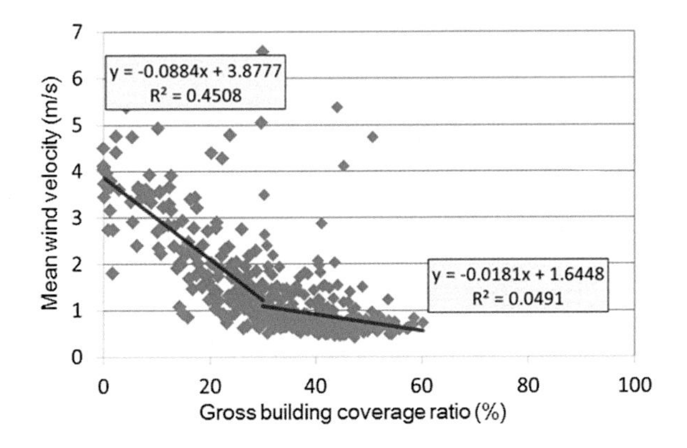

FIGURE 3.37

Relationship between gross building coverage ratio and mean wind velocity at the height of 2 m, averaged over grids 500 m square.

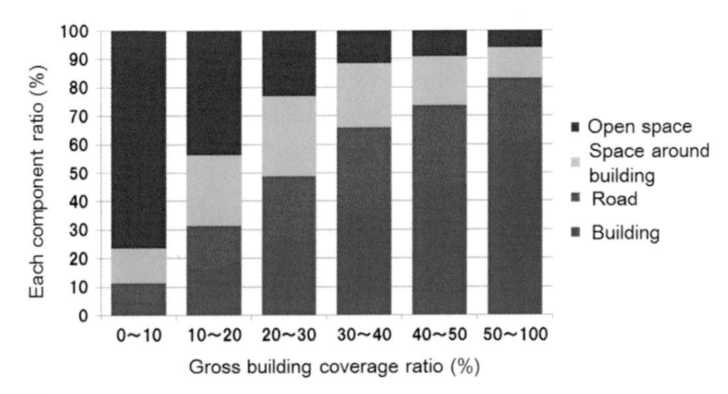

FIGURE 3.38

Relationship between gross building coverage ratio and the other urban block component ratios in grids 500 m square.

velocity is higher over open spaces, such as sea, rivers, and large parks, and lower inside Osaka city where the density of buildings is large.

Urban block components are generally classified as either buildings or open space. In addition, open space is classified as road, space around buildings, and independent open spaces (such as parks or rivers). The classification of urban block components is shown in Fig. 3.35. "Space around building" means the open space belonging to each building site (e.g., approach, garage, plantings).

The relationship between the urban block component ratio and the mean wind velocity averaged over grids 500 m square is shown in Fig. 3.36. Mean wind velocity is better explained by the open space ratio rather than the gross building coverage ratio (building ratio). Relationships between the gross building coverage ratio and mean wind velocity averaged in 500 m grid, and between gross building coverage ratio and the other urban block component ratios in 500 m square grids are shown

Table 3.5 Determination coefficient for each urban block component ratio to mean wind velocity averaged at each evaluation scale.

	2500 m	1250 m	500 m	250 m	100 m
Gross building coverage ratio	0.83	0.79	0.53	0.40	0.16
Road ratio	0.56	0.54	0.31	0.17	0.01
Space around building ratio	0.12	0.01	0.00	0.00	0.01
Open space ratio	0.76	0.74	0.64	0.46	0.16

in Figs. 3.37 and 3.38. When the gross building coverage ratio was less than 30%, the mean wind velocity decreased with the increase of this ratio. In contrast, when the gross building coverage ratio was more than 30%, mean wind velocity and the open space ratio were almost constant. Overall, the mean wind velocity averaged in 500 m square grids is influenced more by the open space ratio rather than by the gross building coverage ratio.

The determination coefficient for each urban block component ratio to the mean wind velocity averaged at each evaluation scale is shown in Table 3.5. Determination coefficients by gross building coverage ratio and open space ratio were as large as, and almost the same in, 2500 m and 1250 m square grids; those by open space ratio

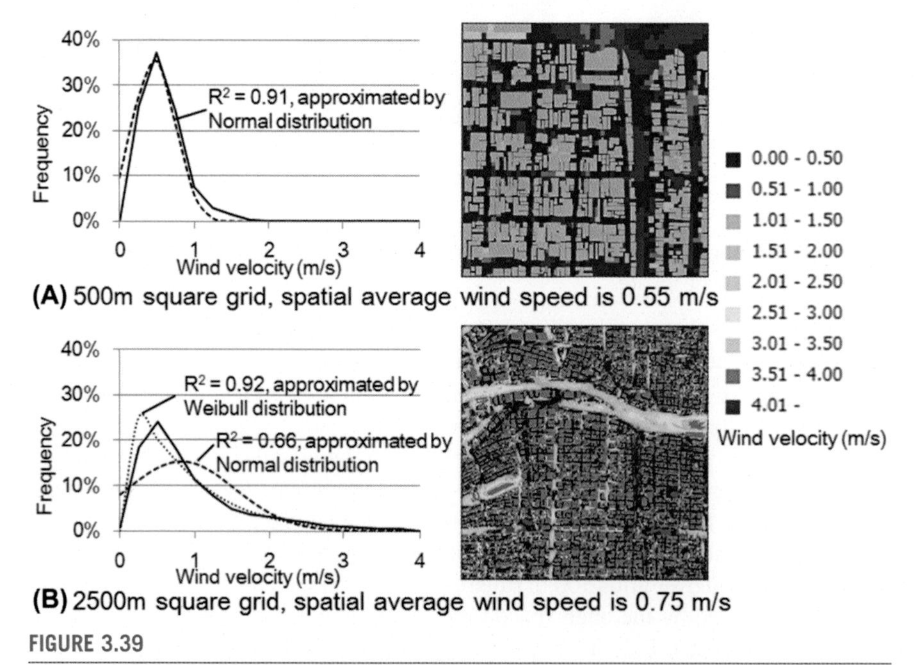

FIGURE 3.39

Frequency distribution of wind velocity at the height of 2 m, and its approximation in grids 500 m (A) and 2500 m (B) square.

Table 3.6 The ratio of areas where the determination coefficient approximated by Normal, Weibull distribution is more than 0.7.

	2500 m	1250 m	500 m	250 m	100 m
Approximated by normal distribution	25%	50%	55%	51%	29%
Approximated by Weibull distribution	75%	81%	83%	78%	37%

were larger than those in 500 m and 250 m square grids. The relationship between any urban block component ratio and mean wind velocity was not confirmed in 100 m square grids. Overall, the mean wind velocity was explained by the open space ratio in grids more than 250 m square.

The frequency distribution of the calculated results and their approximation in 500 m and 2500 m square grids are shown in Fig. 3.39. The accuracy of the approximation by normal distribution in 2500 m square grids was reduced in regions where wind velocity was greater, such as rivers and parks. The ratio of the number of areas where the determination coefficient was approximated by the Normal and Weibull distribution was more than 0.7 (Table 3.6). This ratio was approximated more appropriately by the Weibull distribution. Both methods are inappropriate in 100 m square grids. The ratio of inappropriate areas is a little larger in 2500 m square grids, because there is a possibility that there were two or more wind velocity peaks in the area. Approximation may be worse in the case of larger evaluation areas. After all, the mean wind velocity averaged over an area of about 250 to 1250 square meters is meaningful.

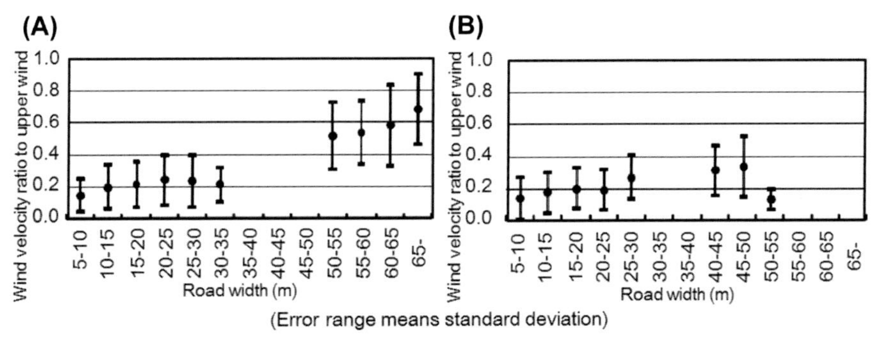

FIGURE 3.40

Relationship between road width and ratio of wind velocity at 2 m to wind velocity at 42 m. (A) For a road parallel to the main wind direction and (B) for a road perpendicular to the main wind direction.

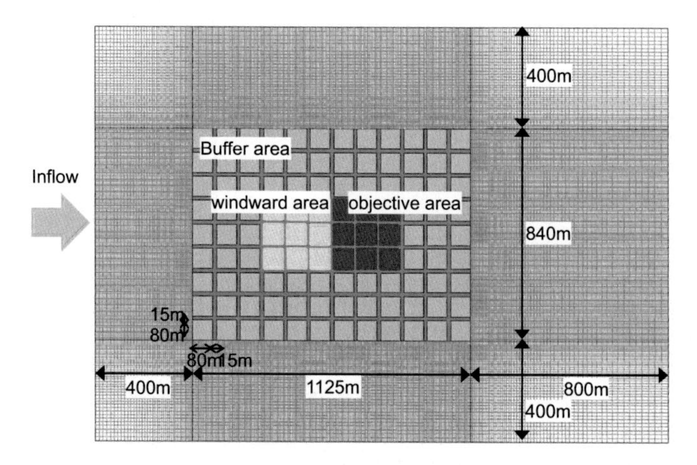

FIGURE 3.41

Setting of the research (objective) and windward areas.

4.2 Possibility of wind environment improvement

The relationship between road width and the ratio of wind velocity at 2 m to wind velocity at 42 m is shown in Fig. 3.40. The left side indicates a road parallel, and right side a road perpendicular, to the main wind direction. As these figures were made from the calculations for the city center of Osaka, there are road widths with no calculation results. Ventilation in street canyons improved on wider roads parallel to the main wind direction. According to the wind environment evaluation scale proposed by Murakami et al. [12], the wind is considered discomforting due to small wind velocity when the latter is less than 0.175 m/s.

A supplemental calculation was carried out using the aligned urban block model with uniform heights of the windward and objective areas. Here, the urban block was 80 m square, road width was 15 m, and building height was changed from 20 to

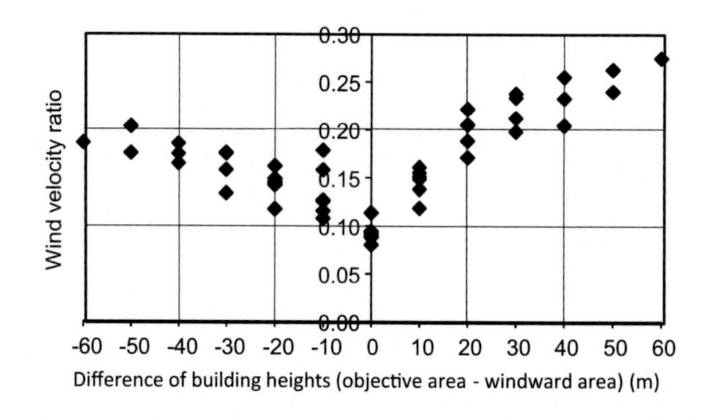

FIGURE 3.42

Relationship between the difference in building heights and mean wind velocity ratio at 2 m, to upper level wind at 54 m.

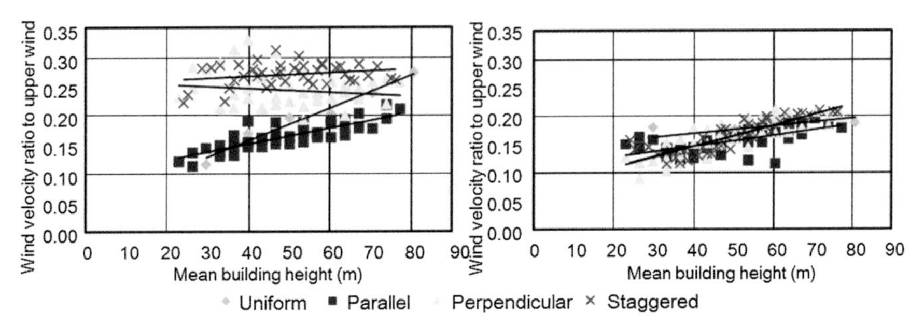

FIGURE 3.43

Relationship between mean building height and mean wind velocity ratio at 2 m, to upper level wind at 54 m (left: building height varies in research area; right: in windward area).

80 m, in every 10 m, in reference to the city center of Osaka. The settings of the research area and the windward area are shown in Fig. 3.41. The relationship between the difference of building heights (between research and windward areas) and the mean wind velocity ratio for upper level wind in the research area is shown in Fig. 3.42. When the difference in building heights was large, the mean wind velocity ratio was large. The mean wind velocity was much higher due to downdrafts when the building height in the research area was higher than that in the windward area. The relationship between the mean building height and mean wind velocity (upper level) ratios is shown in Fig. 3.43. The left side shows results when building height in the research objective area changes; right side is in the case that building height in windward area changes. Buildings arrangement is shown in Fig. 3.44. If there is a variation in perpendicular or staggered arrangement of building heights in objective area, ventilation in the street canyon is improved even if mean building height is low. Influence of building height in windward area on wind velocity in street canyons was not large.

4.3 Selection criteria of hot spot [9]

The probability ratio of occurrence of discomforting wind due to small wind velocity is shown in Table 3.7 for a combined effect of both different road width and

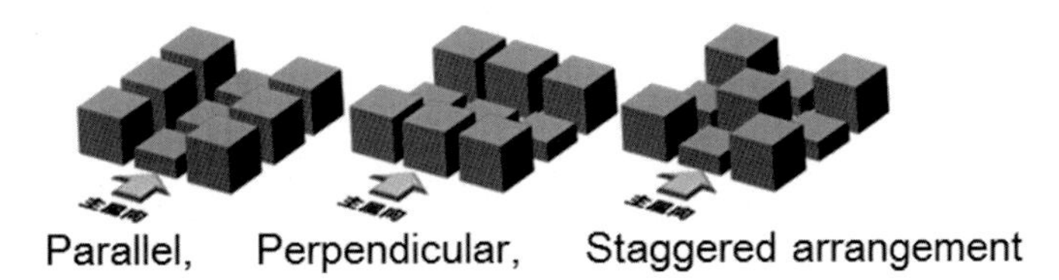

FIGURE 3.44

Building arrangement (left: parallel; center: perpendicular; right: staggered).

Table 3.7 Weak wind risk estimations based on road width and building height.

road width	road parallel to upper wind	road perpendicular to upper wind
0 - 5 m	**74 %**	**78 %**
5 - 10 m	59 %	67 %
10 -15 m	52 %	57 %
15 -20 m	47 %	68 %
20 -25 m	48 %	35 %
25 -30 m	51 %	
30 -35 m		45 %
35 -40 m	0 %	26 %
40 -45 m		27 %
45 -50 m	7 %	**84 %**
50 -55 m	6 %	
55 -60 m	10 %	
60 -65 m	3 %	
65 -70 m	9 %	

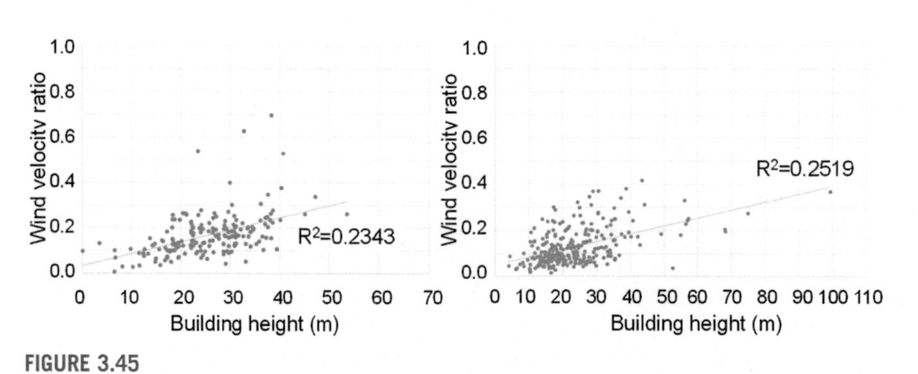

FIGURE 3.45

Relationship between building height and wind velocity ratio. (A) Road parallel to the main wind direction and (B) road perpendicular to the main wind direction.

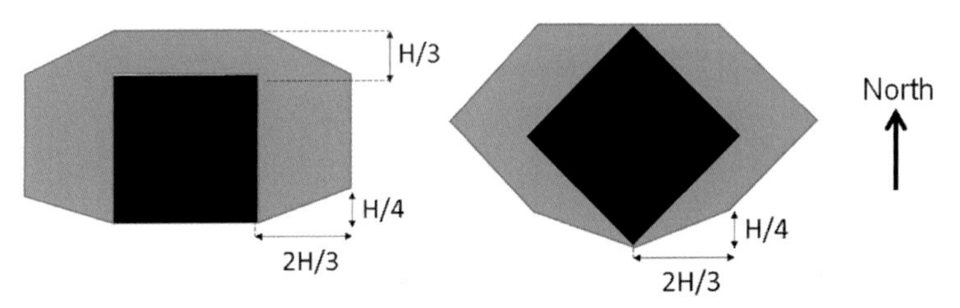

FIGURE 3.46

Selection criteria of hot spot by radiation environment (low-priority area).

Table 3.8 Selection criteria of hot spot by wind environment (high-priority area).

Road width	Road parallel to main wind direction	Road perpendicular to main wind direction
0–5 m	Regardless of building height	
5–10 m	Building height less than 30 m	
10–15 m		Building height less than 40 m

building height. When the risk of occurrence of discomforting wind is more than 70%, the corresponding area is defined as a high weak wind risk area.

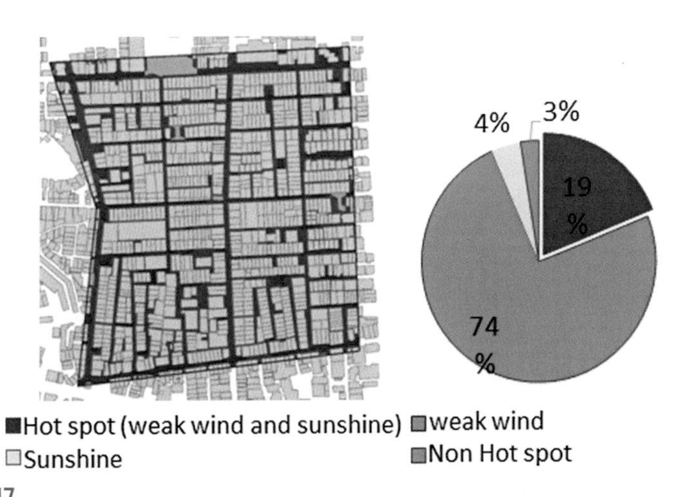

FIGURE 3.47

Hot spot selection result in detached house district.

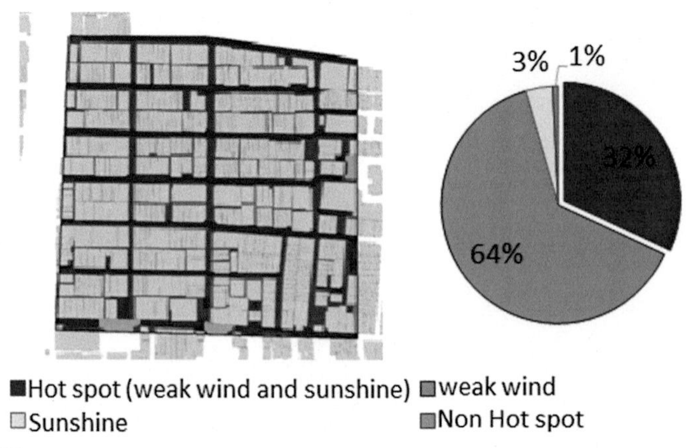

FIGURE 3.48

Hot spot selection result in tenement house district.

Relationship between building height and wind velocity ratio is shown in Fig. 3.45. It is observed that in the low building height case, wind results are discomforting due to small wind velocity both in case of a road parallel to the main wind direction as well as in the case of a road perpendicular to the main wind direction.

If building height is less than 30 m, high weak wind risk areas are defined when the road width is between 5 and 15 m in roads parallel to the main wind direction and between 5 and 10 m in the perpendicular case. If building height is less than

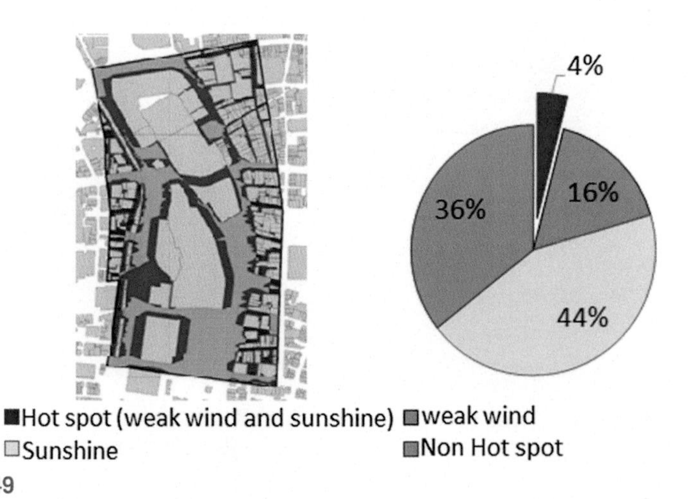

FIGURE 3.49

Hot spot selection result in commerce district.

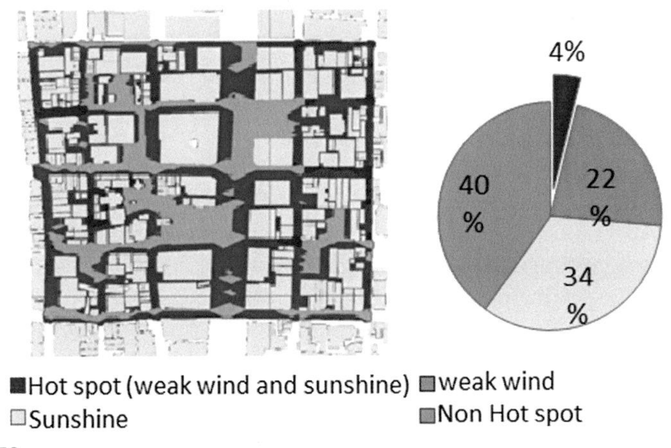

FIGURE 3.50

Hot spot selection result in business district.

40 m, high weak wind risk areas are defined when the road width is between 10 and 15 m in roads perpendicular to the main wind direction.

5. Examples of hot spot selection

Based on the above analysis results, selection criteria of hot spot by radiation and wind environment are shown in Fig. 3.46 and Table 3.8. These criteria are adapted to actual urban blocks using the GIS system.

Hot spots are extracted for typical urban districts in Osaka city. Hot spot selection results in detached house, tenement house, and commerce and business district are shown in Figs. 3.47–3.50. Hot spots with high risk of weak wind and a lot of solar radiation gain are indicated in red color. There are many hot spots in detached house and tenement house districts, whereas less in commerce and business district. As there are a little more open spaces in commerce and business district, it is less risk due to weak wind. Many hot spots are confirmed especially on the north side of the east–west road where the road width is narrow.

6. Summary

Based on the analysis results of the relationship between urban morphology and radiation environment, the followings are clarified. Roof and road surfaces have a higher priority as heat island mitigation locations under general summer conditions. High-priority areas for adaptation measures extend to the buildings on the south sides of the roads for street canyons with larger *W*/*H* ratios on an east–west road.

The distribution is dominated by the shadows of the buildings on both sides of the north–south road. The priority of adopting adaptation measures is low in areas placed at a distance of *H*/3 from the southern building wall and 2*H*/3 from any of the eastern and western walls of the building with a gradient of about *H*/4 toward the northern side.

Based on the analysis results of the relationship between urban morphology and wind environment, the followings are clarified. The mean wind velocity averaged in 500 m square grids is influenced more by the open space ratio rather than by the gross building coverage ratio. A high weak wind risk area is defined in the three following cases: (1) the road width is 0–5 m; (2) a road parallel to the main wind direction has a width between 5 and 15 m, or a road perpendicular to the main wind direction has a width between 5 and 10 m, and the building height is less than 30 m; and (3) a road perpendicular to the main wind direction has a width between 10 and 15 m, and building height is less than 40 m.

From examples of hot spot selection, many hot spots are confirmed especially on the north side of the east–west road where the road width is narrow. The number of hot spots in detached house and tenement house district is larger than that in commerce and business district, due to the amount of open space related to weak wind risk.

References

[1] Ministry of Environment of Japan. Guidelines for measures against heat in the town. 2016 (in Japanese).

[2] Baumueller J, Hoffmann U, Reuter U. Climate booklet for urban development. Ministry of Economy Baden-Wuerttemberg, Environmental Protection Development; 1992.

[3] Ren C, Ng E, Katzschner L. Urban climatic map studies: a review. International Journal of Climatology 2011;31:2213–33.

[4] Kitao N, Moriyama M, Nakajima S, Tanaka T, Takebayashi H. The characteristics of urban heat island based on the comparison of temperature and wind field between present land cover and potential natural land cover. In: Proceedings of Seventh international conference on urban climate; 2009.

[5] Takebayashi H, Moriyama M. Relationships between the properties of an urban street canyon and its radiant environment: introduction of appropriate urban heat island mitigation technologies. Solar Energy 2012;86(9):2255–62.

[6] Takebayashi H, Kimura Y, Kyogoku S. Study on the appropriate selection of urban heat island measure technologies to urban block properties. Sustainable Cities and Society 2014;13:217–22.

[7] Takebayashi H. High-reflectance technologies on building facades: installation guidelines for pedestrian comfort. Sustainability 2016;8:785.

[8] Mino tile commercial cooperative. Mino cool island tile. 2011 (in Japanese), http://minotile.com/about/index3.html. [Accessed 18 December 2014].

[9] Takebayashi H, Kiyama Y, Yamamoto N. Analysis of wind and radiant environment in street canyons for production of urban climate maps at district scale. Journal of Heat Island Institute International 2017;12(2):78–83.

[10] Takebayashi H, Oku K. Study on the evaluation method of wind environment in the street canyon for the preparation of urban climate map. Journal of Heat Island Institute International 2014;9(2):55–60.

[11] Tominaga Y, Mochida A, Yoshie R, Kataoka H, Nozu T, Yoshikawa M, Shirasawa T. AIJ guidelines for practical applications of CFD to pedestrian wind environment around buildings. Journal of Wind Engineering and Industrial Aerodynamics 2008;96: 1749–61.

[12] Murakami S, Morikawa Y. Criteria for assessing wind-induced discomfort considering temperature effect. Journal of architecture, planning and environmental engineering, AIJ 1985;358:9–17 (in Japanese).

CHAPTER

Case studies of adaptation cities

4

Noboru Masuda, Dr.[1], Nobuya Nishimura, Dr.[2], Minako Nabeshima, Ph.D[3], Toru Shiba[4], Hiroyuki Akagawa, Ph.D[5]

[1]*Professor Emeritus, Director of Plant Factory Research and Development Center, Osaka Prefecture University, Sakai, Japan;* [2]*Professor, Department of Mechanical and Physical Engineering, Osaka City University, Osaka, Japan;* [3]*Professor, Department of Urban Design and Engineering, Osaka City University, Osaka, Japan;* [4]*Manager, Residential Energy Business Unit, OSAKAGAS CO.,LTD., Osaka, Japan;* [5]*Senior Engineer, Technical Research Institute, Obayashi Corporation, Tokyo, Japan*

Chapter outline

Adaptation Measures for Urban Heat Islands. https://doi.org/10.1016/B978-0-12-817624-5.00004-X

1. Characteristics of cool spots and cool roads in big city Osaka

1.1 Characteristics of Osaka Prefecture

As Japan's second largest metropolis after Tokyo, Osaka Prefecture holds the third largest population of about 8.82 million, following Tokyo and Kanagawa, and boasts the second largest gross regional product after Tokyo. Fig. 4.1 indicates the location of Osaka in Japan, and Fig. 4.2 [1] shows the current situation of land use in Osaka. The prefecture is surrounded by Hokusetsu, Ikoma, and Izumi-Katsuragi mountains, with its urban area widely spreading over the central plain.

Situated at a mid-latitude region, from 34 degrees°16′ to 35 degrees°03′ north latitude, and at 135 degrees°05′ to 44′ east longitude, Osaka belongs to a warm-temperate zone. At altitudes ranging from the sea level to 959 m, with its annual average temperature being 16.5°C and its average rainfall about 1300 mm, Osaka has 68 scorching days per year, the nation's second largest in number of such days on which the temperature exceeds over 30°C.

Osaka has been warming rapidly, as its temperature has risen by 2.1 degrees over the last 100 years, which is 1.1 degrees higher than the national average of 1.0 degrees. The higher speed of warming is believed to be caused by the heat island effect. In Osaka, the number of scorching days has increased 1.4 times over the last three decades. Fig. 4.3 [2] shows the temperature distribution in 2004, in which the high-temperature zones spread in urban areas in the central plain, with low-temperature zones lying along the peripheral Ikoma, Kongo, and Izumi-Katsuragi mountains, except for Hokusetsu mountains corresponding to the mid-temperature zone.

In these circumstances, the Osaka prefectural government has formulated a plan for promoting measures to mitigate heat island effect in Osaka, aiming to build a

FIGURE 4.1

Location of Osaka.

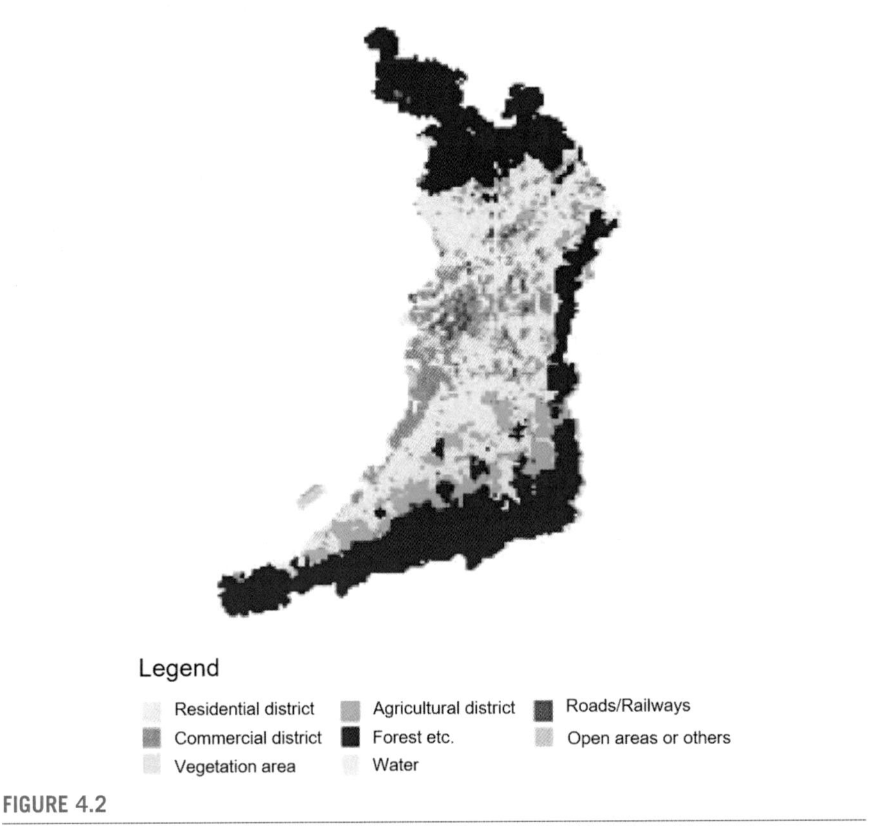

FIGURE 4.2

Land use situation of Osaka Prefecture in 2000.

heat-island-effect—conscious city. Under the plan, it set targets for (1) reducing the number of sweltering nights by 30% from the current level by 2025 by lowering the summer night temperatures in residential areas, and (2) lowering perceived temperatures by creating outside cool spots in an effort to mitigate the daylight summer heat environment.

In 2012, the Osaka Heat Island Countermeasure Technology Consortium, known as HITEC, launched a campaign for raising awareness about cool places by conducting a survey to select Osaka's best 100 cool spots, in an attempt to help alleviate a feeling of discomfort caused by urban heat island phenomena and to increase comfortable outdoor spaces. In 2015, it further expanded its selection category to the best 100 cool roads, to enhance recognition of linear collections of cool spots and encourage the usage as a way of helping cope with severe summer heat. In contrast to cool spots that deliver the information of their own single location, cool roads, each connecting such cool spots or dots, can provide more information along their courses, including commuting paths, walkways, and shopping routes.

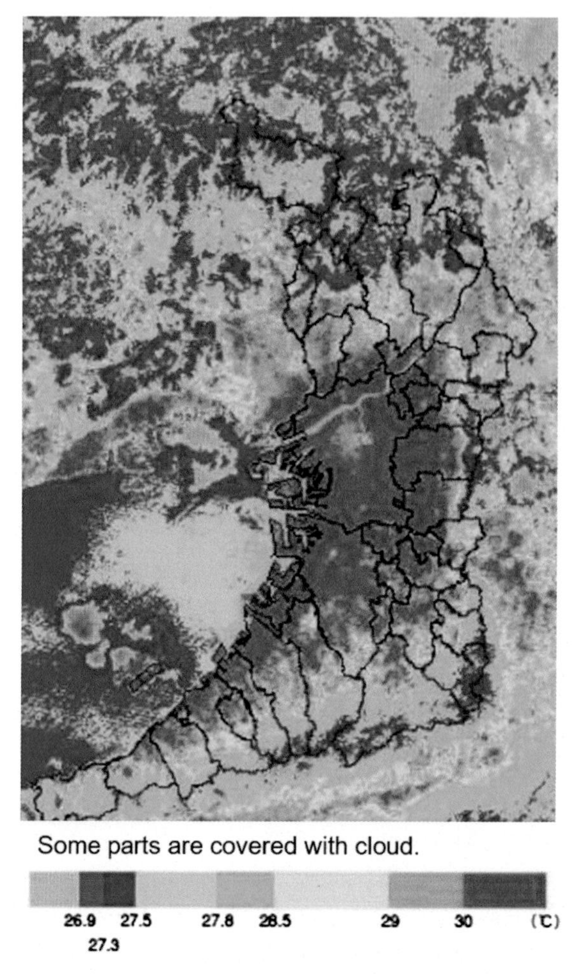

FIGURE 4.3

Temperature distribution in and around Osaka Prefecture at 10:00 a.m. on 12 August in 2004.

1.2 Cool spots

Within Osaka Prefecture, 119 locations have been recognized as cool spots or cool roads, as shown in Fig. 4.4 based on the data of Google Map [3]. Many of them are widely distributed in Osaka City holding 37 locations, as well as in the Hokusetsu area with 38, followed by Higashi-Osaka having 20, Minami-Kawachi 8, and the Senshu region 16.

Fig. 4.5 shows areas capable of forming cool spots by kind, while Fig. 4.6 indicates environmental factors that survey respondents consider as constituting cool spots.

By kind, among areas capable of forming cool spots, forests or natural parks make up the largest portion of 26.1%, followed by city parks in urban areas

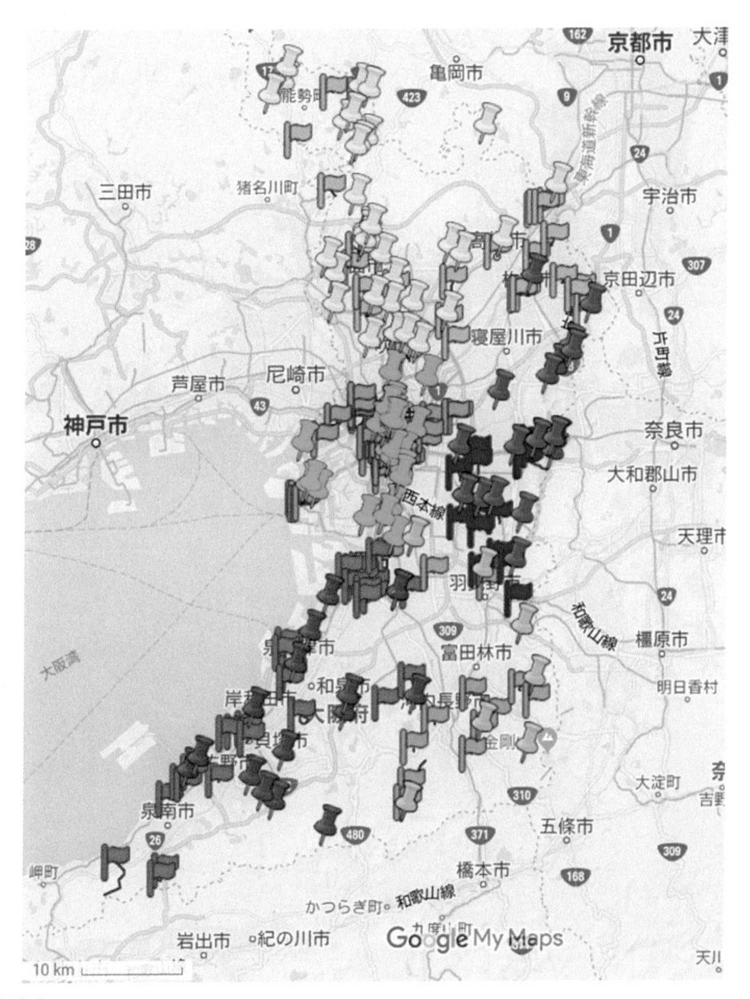

FIGURE 4.4

Locations of cool spots (pin marks) and cool roads (flag marks).

Map data from ©2019 Google.

representing 22.7%. The third largest contributor is clear spaces adjacent to and required by buildings in city centers with 16.8%, followed by temples and shrines with greenery accounting for 13.4%. Linearly extended places such as pathways, tree-lined streets, and shopping arcades each comprise a small percentage, less than 10%, which signals that there is still not so much coolness people can feel on the move.

As to environmental factors, the single factor of shade of trees dominantly takes up the largest ratio of 28.6%. It is followed by water-related factors: water misting and sprinkling in the second place with 9.2%, and water surface taking up 8.4%,

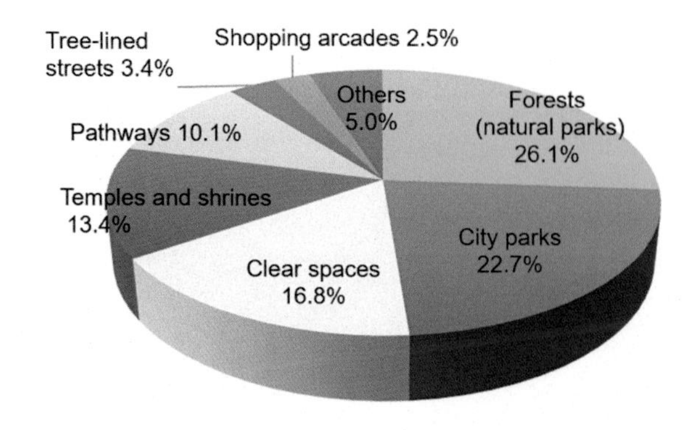

FIGURE 4.5

Cool spot forming factors.

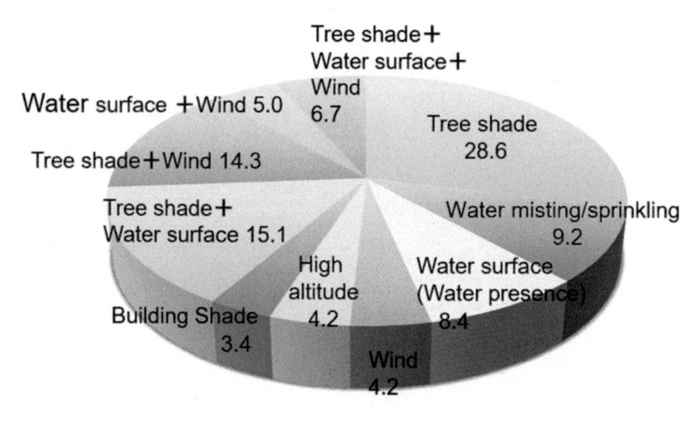

FIGURE 4.6

Cool spot environmental factors.

which signifies the presence of water matters a lot. When we look at mixed environmental factors, the combination of tree shade and water surface and that of tree shade and wind represent 15.1% and 14.3%, respectively, meaning the presence of trees that block sunlight plays a key role in forming cool spots as well. Here again, the presence of water cannot be ignored in that an evaporative cooling effect can be expected.

Fig. 4.7 [4] illustrates the details of the abovementioned typical cool spots: natural parks, city parks, and clear spaces, as well as temples and shrines. These radar charts indicate the intensity scale and differences of environmental factors, comprising greenery, water, wind, sound, shade, and flowers at each cool spot. In addition, since tree shade has a great influence on coolness, the photos of sky view factors are included here, showing their visible sky percentages are roughly less than 40% in all. Interestingly, the presence of flowers or sound, seemingly unrelated to ambient temperature changes, is noted as factors stimulating a sense of coolness.

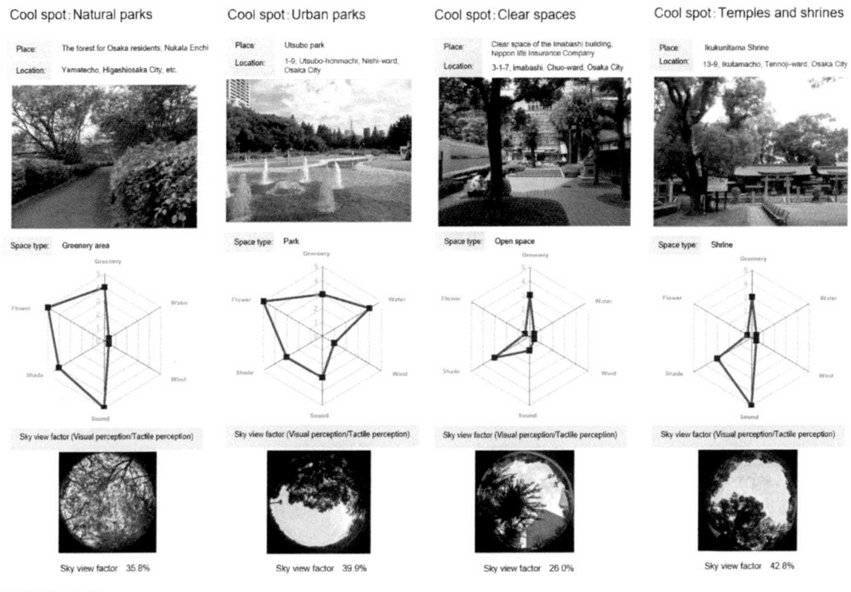

FIGURE 4.7

Evaluation of typical cool spots.

1.3 Cool roads

A total of 121 cool roads in Osaka Prefecture have been chosen, as shown in Fig. 4.4. Osaka City holds the largest number of locations (36), followed by the Hokusetsu area with 25, Higashi-Osaka 15, and Minami-Kawachi 14. Unexpectedly, the Senshu area has as many as 31 such locations.

Fig. 4.8 describes areas capable of forming cool roads, whereas Fig. 4.9 indicates environmental factors that survey respondents consider as constituting cool roads.

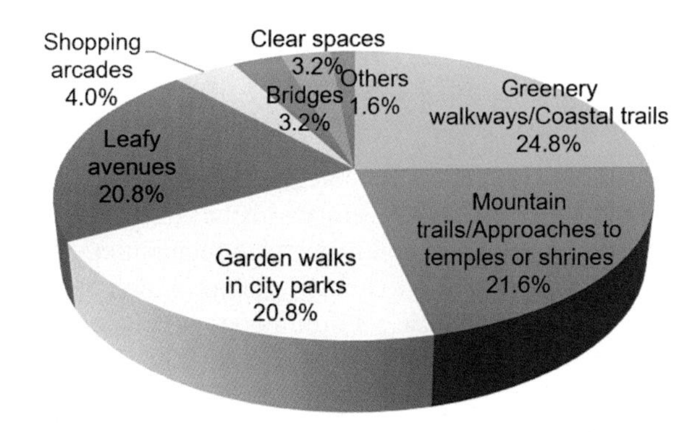

FIGURE 4.8

Cool road forming factors.

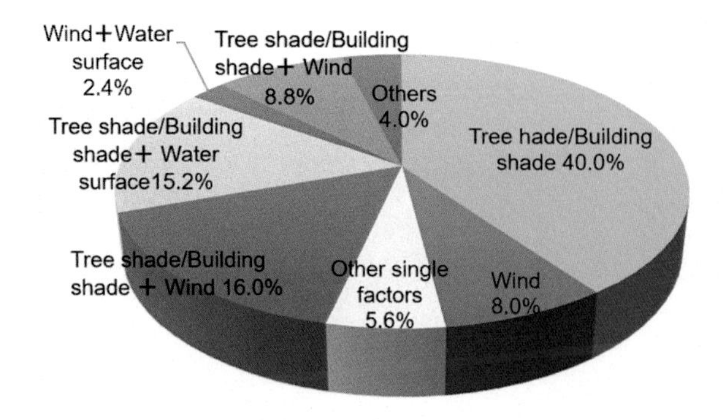

FIGURE 4.9

Cool road environmental factors.

The cool road is roughly defined as an over-50-meter stretch of cool space located near a workplace or living place. People in Osaka offered various candidates meeting this requirement. Among those capable of becoming cool roads, waterfront greenery walkways or coastal trails account for the largest number of 24.8%. Mountain trails or approaches to temples or shrines in peripheral mountains come second at 21.1%, followed by garden walks in city parks at 20.8% and leafy avenues in urban areas at 20.8% as well. Other components make up relatively small portions.

In terms of environmental factors, similarly to the cool spot constituents, the biggest is a single factor of the shade of trees or buildings with 40.0% followed by wind with 8.0%. As mixed or multiple factors, the combination of tree or building shade plus wind accounts for 16.0%, while that of the shade plus water surface represents 15.2%. As in the case of cool spots, the presence of shade serves as an indispensable cooling factor in forming cool roads. Intriguingly, however, linearly shaped cool roads are found to be more largely affected by winds blowing across water surfaces or between buildings.

2. Countermeasures by using water

2.1 Novel water facilities for creation of comfortable urban micrometeorology

In Japan, high air temperature experienced in urban areas has become a major issue, termed the "heat island phenomena." Together with the effects of the blazing sun and the heat radiating from surrounding buildings, these high air temperatures make the thermal environment in urban spaces extremely uncomfortable in summer. The provision of trees, lawns, and applied water are important countermeasures that ease this environment a bit. The cooling effects of trees and lawns have been well researched and many reports are available [5,6], and now we can see many trees and lawns in urban areas.

Conventional water facilities, such as fountains, sprays, and canals, have already been installed in urban areas. But this has been mainly for landscaping purposes, so there has been no attempt made to quantify and capitalize on their cooling effect. Therefore, we tested to place water facilities in rather narrow city spaces, such as beside sidewalks and between buildings [7]. The aim is to make the most efficient use of water's ability to adjust temperature and humidity, as well as providing its landscaping and hydrophilic functions. By doing so we have contrived a new facility that can be adjusted according to the micrometeorological environment in the living space.

For this purpose, we carried out a field study on the degree of cooling effect that artificial water facilities can produce in relatively narrow urban areas.

2.1.1 A proposal for water facilities for environmental creation

We proposed new water facilities for environmental control functions in addition to the existing functions as shown in Fig. 4.10. The concept is to position water facilities as a countermeasure of forming a comfortable urban environment. Size, types, and layout of the water facilities can be determined based on the area of urban space to be controlled, as shown in Table 4.1.

With respect to the effects of water to establish facilities in the macro-space on environment creation, extensive research has been made from a passive viewpoint of the natural cooling effects of rivers and ponds to the neighboring environment. But there has been no such research so far from the stance of positive utilization of water facilities.

Hereupon, we tried to propose novel water facilities for environment creation in the meso-space from the basic stance of creating comfortable living and moving spaces for people including both in and out of buildings. A concept image for this water facility placed in the meso-space is shown in Fig. 4.11. A cool spray hanging

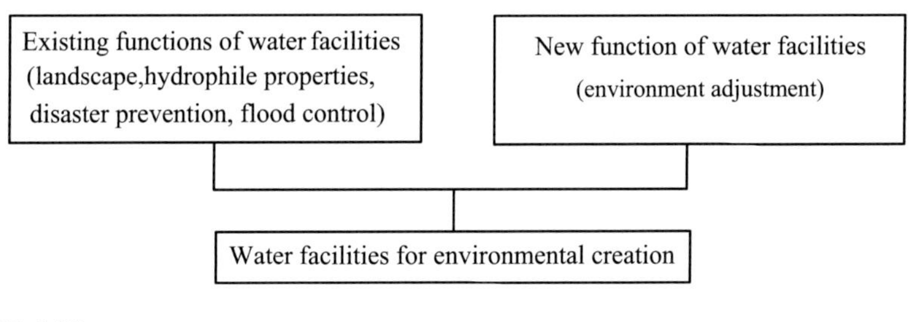

FIGURE 4.10

Concept of water facilities for environment creation.

Table 4.1 Functions of the artificial water facilities for environment creation.

1	Environment creation of macro-space To control the macro-space environment by placing water facilities throughout a whole city
2	Environment creation of meso-space To create environment in the limited zone of meso-space by cooling-spots or zones placed in open space around building or pavement
3	Environment creation of micro-space Conventional environment control for closed spaces such as inside room

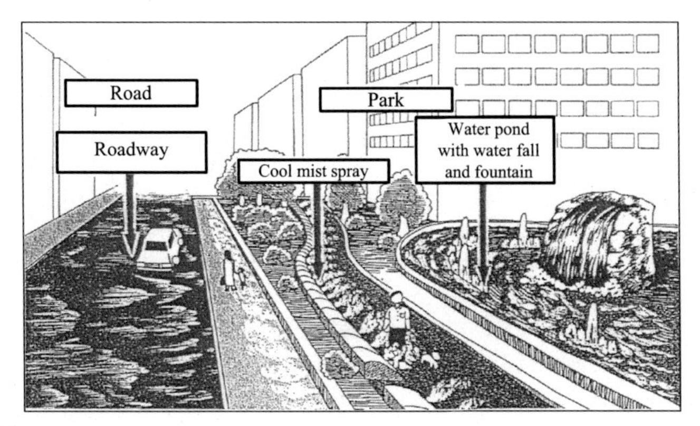

FIGURE 4.11

Image of the novel water facilities for environment creation.

in the air around the hot sidewalk makes people feel comfortable. In addition, a fountain has been placed there.

2.1.2 Field measurements and a proposal for urban water facility

With an aim at proposing an attempt to improve micrometeorologic thermal environment as mentioned above, we next used the existing water facilities to investigate the possibility of temperature-humidity adjustment and required conditions to establish facilities in the field. As an existing water facility, an artificial water facility in a park in an urban area (Tennoji Park in Osaka city, a typical big city in Japan) was used. There are four characteristic water facilities in this park. We focused on the pond (called, "Stage Stone Pillar Pond," about 0.2 m deep) which combines a waterfall on the side of stone pillars and a spray fountain as shown in Fig. 4.12. Measurement points were set around the pond at 60 m east and west, and 40 m north and south. Fig. 4.13 shows temperature measurement points.

FIGURE 4.12

Stage Stone Pillar Pond at Tennoji park.

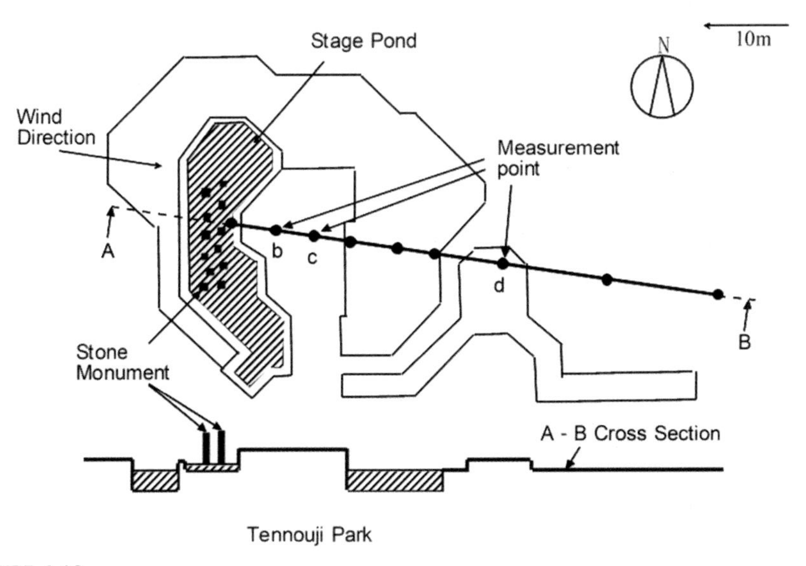

FIGURE 4.13

Temperature measurement points around the water facility.

Measurements took place on July 30 and 31, 1992, when summer conditions were at an average level. During the two days of the measurement, the sea breeze of 2.5 m/s west-northwest was dominant in clear weather. There are 25 stone pillars in the Stage Stone Pillar Pond and the pond was on average 2.0 m high, 0.5 m wide, and 0.2 m deep. A waterfall from the top of these stone pillars and a spray fountain spouting water about 4 m wide from near water surface of the pond are repeatedly operated. The fountain consists of 120 spray nozzles, and the amount of water spouted from each nozzle is 1/3 L/second (20 L/min). The waterfall and fountain are operated intermittently from 9:00 to 21:00 with the following pattern:

1. waterfall: 10 min operation at every 20 min
2. fountain: 3 min operation when waterfall is stopped

2.1.3 Measurement results and discussion

2.1.3.1 Temperature changes around the water facility

Fig. 4.14 shows an example of temperature change of air, ground surface, and water at representative measuring points around the Stage Stone Pillar Pond for 1 day. On the paved stone, the maximum ground surface temperature reaches above 50°C in the daytime, but the water temperature in the Stage Stone Pillar Pond is around 30°C, and the maximum ground surface temperature on the wet paved stone by the spray fountain was about 39°C. Furthermore, the nighttime ground surface temperature on the paved stone was substantially higher than the air temperature due to the stored heat effect. On the other hand, we can observe that the ground surface temperature on the wet paved stone in the Stage Stone Pillar

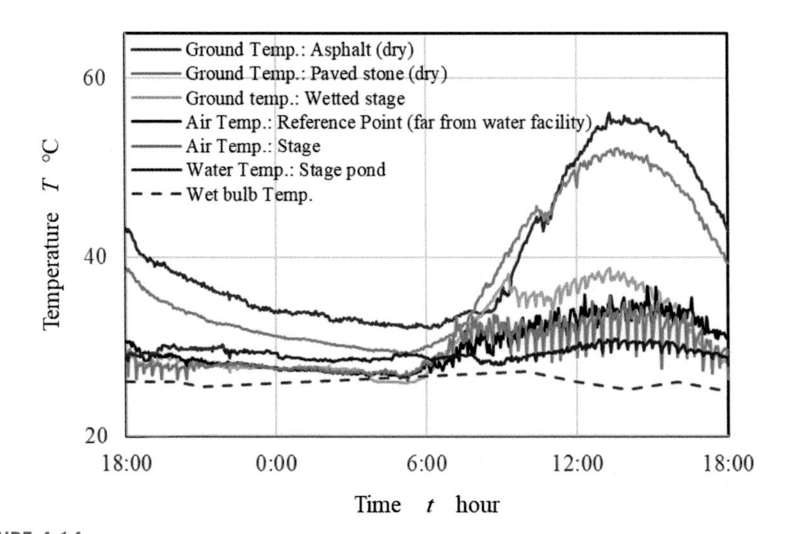

FIGURE 4.14

Temperature changes around the water facility.

Pond went down as low as the air temperature in the evening because of evaporation and cooling of the moisture. This enables us to predict that we can adjust the lowering range of air and ground surface temperature by controlling the diffusing area of the fountain.

2.1.3.2 Air temperature changes on the leeward side

Fig. 4.15B–D shows air temperature changes and the cooling effects measured during spray operation from 12:00 to 15:00 on July 31 at the representative measuring points b, c, and d on the leeward side of the Stage Stone Pillar Pond. In the graph (A) of Fig. 4.15, bars show operating times of the spray. It can be observed from these results that the closer the measuring points are to the stage pillars, the more the effect of temperature fall by the spray can be obtained. These degrees of temperature decline and fluctuation seemed to be affected by elements such as wind direction, velocity, and turbulence.

2.1.3.3 Distribution of average temperature and humidity on the leeward side

Fig. 4.16 shows the distribution of average temperature and humidity on the leeward side from 14:00 to 15:00. Similarly to Fig. 4.15, the closer the measuring points are to the Stage Stone Pillar Pond, the lower the temperature went down during spray operation. Conversely, humidity was higher when the spray was operated. These results tell us that the temperature fall area induced by evaporation of water from the spray and waterfall operation at the Stage Stone Pillar Pond spreads out to a distance of nearly 35 m toward leeward from the Stage Stone Pillar Pond. Also, even when the spray was not operated, air temperature became 1–2 K

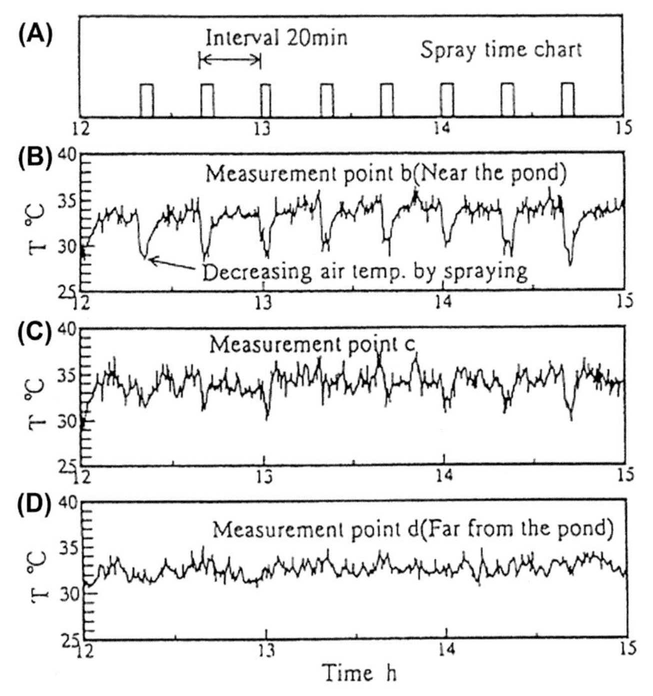

FIGURE 4.15

Air temperature changes on the leeward side.

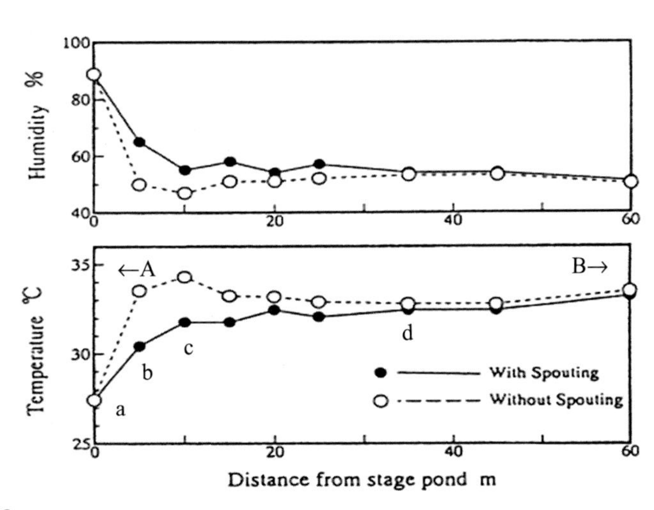

FIGURE 4.16

Distribution of the average temperature and humidity of the air current above the line between points A and B.

lower than the average air temperature in this park, that is, 35°C. Thus, it was confirmed that spray or waterfall operations create an air temperature decline area on the leeward side, and that the degree of temperature decline depends on types of water facilities, but especially water spray facilities are effective for cooling outdoor air.

2.2 District heating and cooling system using thermal energy of river water

2.2.1 System overview

Nakanoshima located at the center of Osaka City is an area having rich water resource surrounded by Dojimagawa River and Tosaborigawa River as shown Fig. 4.17.

The district heating and cooling (DHC) system that uses unutilized energy of river water was constructed [8] in December 2004 at Nakanoshima 3-chome, which is the rectangle area in the figure. The DHC system was planned to effectively use of the unutilized thermal energy of river water. Water is taken from Dojimagawa River and drained off into Tosaborigawa River by capitalizing on the geographic features of Nakanoshima as shown Fig. 4.18 [9].

Heat pumps can be more efficiently operated by using river water than air-source heat, because the temperature of river water is lower than that of the air during summer and higher than that of the air during winter as shown Fig. 4.19.

Table 4.2 shows the specifications of main thermal supply system that consisted of heat pumps and ice storage tanks.

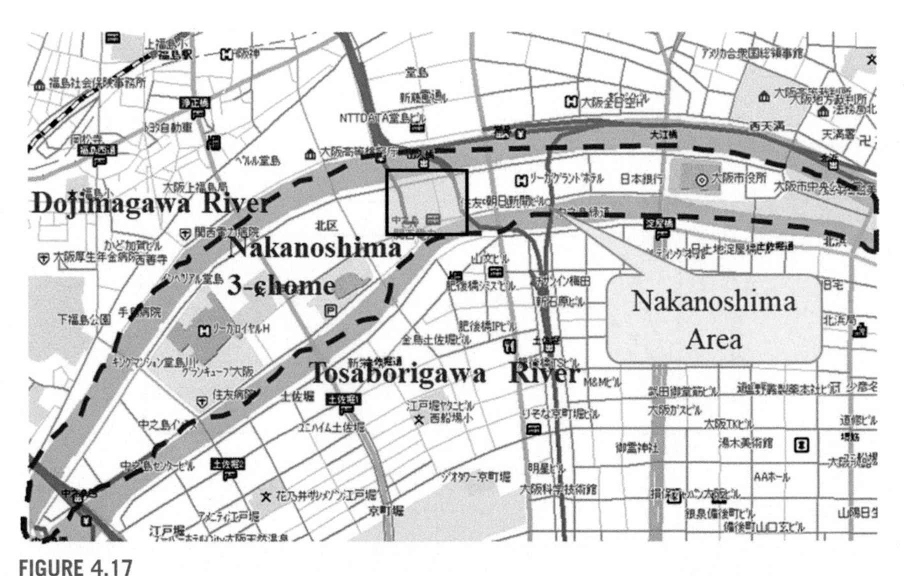

FIGURE 4.17

Location of the Nakanoshima 3-chome district heating and cooling system.

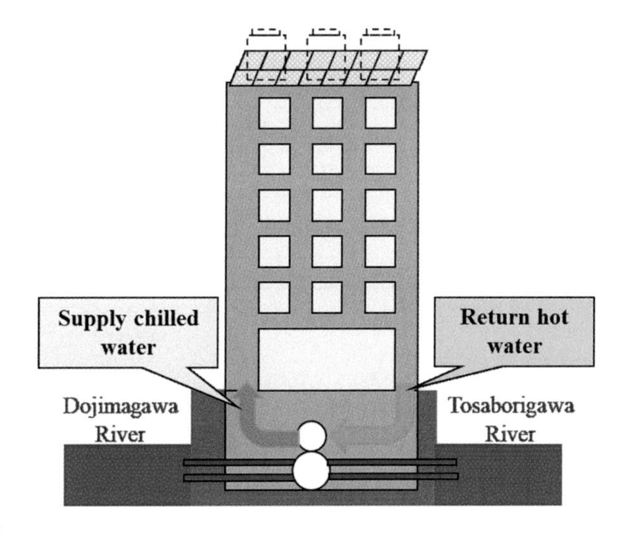

FIGURE 4.18
Outline of the river water utilization in district heating and cooling system.

2.2.2 Evaluation of sensible exhaust heat reduction to the air using thermal energy of river water

Fig. 4.20 shows the yearly change of the production heat quantity of cooling/heating energy from the start of plant operation [10]. The figure shows that cooling demand is more than three times the heating demand. In addition, both cooling demand and heating demand have increased since 2009. This is because additional heat source equipment was installed in 2009, as shown in Table 4.2.

Then, the verification results on the heat island mitigation effect of the heat pump driven district heating and cooling system utilized river water that based on secular operation data are described.

Definition of Q_{LHI} waste heat from building to atmosphere

$$Q_{LHI} = Q_o - Q_i$$

where Q_i is the amount of heat absorbed from the atmosphere and Q_o is the amount of heat exhausted to the atmosphere.

The heat balance of the virtual closed space surrounding the building is expressed by the following equation.

$$E_b + E_d + Q_i + Q_m = Q_o + Q_d + Q_w + (Q_r - Q_s)$$

Here, E_b, E_d, Q_m, Q_d, Q_w, Q_r, and Q_s mean the following power consumption and heat dissipation.

E_b: Building power consumption
E_d: DHC system's power consumption
Q_d: Heat dissipation to river
Q_m: Heat generation amount by the human body

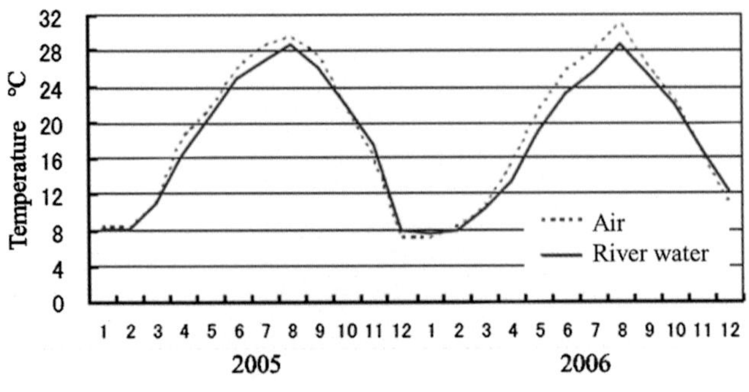

FIGURE 4.19

Seasonal changes of the air temperature and river water temperature.

Table 4.2 List of installed heat source equipment for each construction period.

1st stage			
	Cooling	Heating	Number
HP	-	838 MJ/h	1
IHP	Cool water : 3,080 MJ/h Ice Storage : 1,936 MJ/h	Cool water heat recovery : 3,606 MJ/h Ice Storage heat recovery : 2,448 MJ/h	8 Unit (16)
TR1	5,063 MJ/h	-	1
	Storage capacity		Number
IST1	Dynamic type 139,440 MJ 870 m3		8
2nd stage			
	Cooling	Heating	Number
SR1	Cool water : 5,062 MJ/h Ice Storage : 4,404 MJ/h	4,187 MJ/h	1
SR2	Cool water : 8,640 MJ/h Ice Storage : 8,478 MJ/h	13,860 MJ/h	1
TR2	7,595 MJ/h	-	1
	Storage capacity		Number
IST2	Static type 78,230 MJ 545 m3		8 Unit
3rd stage			
	Cooling	Heating	Number
R31/R32	8,561 MJ/h	8,910 MJ/h	2
	Cooling	Heating	Number
HWR41/ HWR42	-	464 MJ/h	2
HSR43	-	184 MJ/h	1
HSR44	113 MJ/h	184 MJ/h	1

TR, Water Cooling Turbo Chiller; IHP, Water Source Screw Heat Pump (Ice storage and heat recovery); IST, Ice storage tank; SR, Water Source Screw Heat Pump (Ice storage and change of cool and warm water mode; HP, Water Source Screw Heat Pump (Heating); R, Water Source Screw Heat Pump (change of cool and warm water mode); HWR, Scroll Heat Pump (high temp.); HSR, Scroll Heat Pump (high temp. and heat recovery)

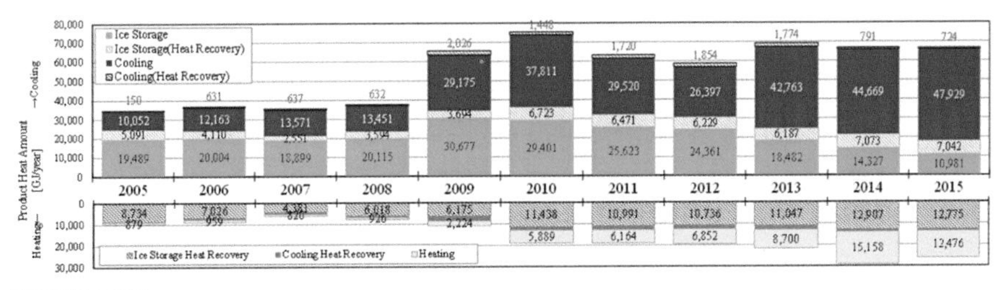

FIGURE 4.20

Amounts of heat for space cooling and space heating that were produced by the district heating and cooling plant.

Q_r: Waste heat
Q_S: Amount of heat storage
Q_W: Hot water exhaust heat

Therefore, building's heat load Q_{LHI} for heat island is calculated in the following equation.

$$Q_{LHI} = (E_b + E_d + Q_m) - (Q_d + Q_w) - (Q_r - Q_s)$$

Fig. 4.21 shows the estimation results of the effect of reducing the amount of sensible heat exhaust to the atmosphere that causes heat islands by the DHC system using river water. This figure shows the result on August 19, the peak summer day of 2014. The figure shows that the amount of heat exhausted to the atmosphere throughout the day was −83 GJ, indicating that the building absorbed heat from the atmosphere. From the above results, it was demonstrated that the DHC system using river water is extremely effective as a heat island mitigation countermeasure.

3. Case studies using green shade and greening activities

3.1 Case studies

3.1.1 Attempt to mitigate the urban heat effect in an experimental residential complex, called NEXT 21

3.1.1.1 The outline of the NEXT 21 project

NEXT 21 is the name of a housing complex in Osaka City experimentally built by a company to explore what neo-futuristic collective housing should be like (Fig. 4.22). The case study provides a variety of proposals to realize a rich living environment in harmony with nature and reduces the environmental burden in a series of multifaceted fields related to energy, environment, urban planning, architecture, and facilities. Experiments were performed.

After the building construction was completed in October 1993, and later made open to the public, 16 households chosen among employees of the company participated in a habitation experiment starting in April 1994, and have since resided there actually.

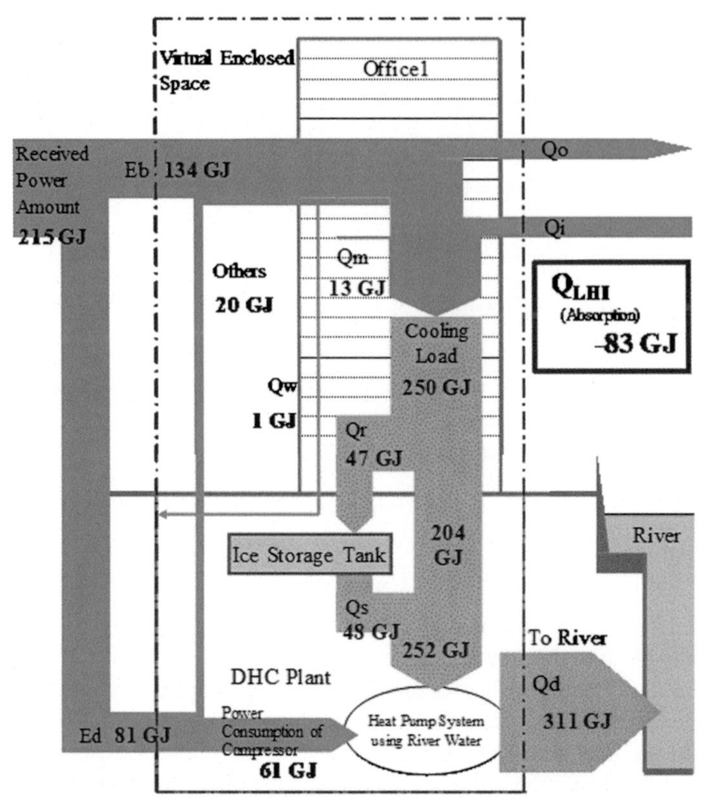

FIGURE 4.21

Exhaust heat load reduction to the air causing heat island using thermal energy of river water. *DHC plant*, district heating and cooling plant.

The building has demonstrated great results in reducing environmental burdens, by cutting back on primary energy consumption thanks to its insulation capability and energy-saving mechanisms including cogeneration equipment, as well as by improving the environment through greening the building and garbage disposal/wastewater treatment systems.

3.1.1.2 Efforts for greening the building

Although having Osaka Castle Park and Tennoji Park, Osaka City is said to be poor in greenery, with its green space with trees or forests accounting for merely about 5% of the total city area. Therefore, efforts for roof afforestation and wall surface greening are believed to be quite significant in expanding urban greenery at the private level.

The NEXT 21 building offers rich greenery not only on the rooftop but on the terraces and verandas, forming vertically connected green extensions on its external surfaces. Since the vast Osaka Castle park is situated 1 to 2 kilometers to the north of

FIGURE 4.22

Appearance of NEXT 21.

the building, an ecology investigation was made into flora and birdlife in green areas around the site before the building construction began, to make the building's forestation plan best suited to the surrounding environment.

After the completion of the building, another 5-year investigation plan was implemented, which proves that the building was visited by 22 kinds of wild birds, and that some of them were observed moving along the vertical green coverage. Also, trees planted in the NEXT 21 premises were found to have contributed to the improvement of its thermal environment.

Fig. 4.23 shows two images of the NEXT 21 building with its surroundings, both taken at around 3:00 p.m. on August 12; Fig. 4.23A taken with visible light, and Fig. 4.23B with infrared light.

While the concrete walls of the buildings around the NEXT 21 facility were extremely heated up to over 40°C against the ambient temperature of 34°C, the temperature of the green areas of the NEXT 21 building stayed at 35°C, almost the same as the air temperature. Furthermore, as shown in Fig. 4.24, the courtyard of the NEXT 21 facility was recognized as forming a large cool spot, which suggests that a cool breeze may likely blow through the building [11].

3.1.1.3 Efforts for energy saving

The NEXT 21 building has demonstrated a greater heat insulation capacity to cut energy consumption. Also, cogeneration systems have been adopted there, initially a central heating system using 100-kilowatt phosphoric acid fuel cells (PAFCs), which was later replaced with an individually installed heating system using solid

(A)

(B)

FIGURE 4.23

Photograph (A) and thermograph (B) of NEXT 21 and its surrounding.

(Photography: Professor Emeritus Akira HOYANO).

oxide fuel cells (SOFCs) that enables highly efficient generation of electricity powered by city gas. At the same time, as the new system allows exhaust heat produced during the power generation to be used for air-conditioning or heating purposes, it is expected to further contribute to energy saving. Fig. 4.25 is solid oxide fuel cells-based Cogeneration system.

In addition, a photovoltaic generation system has been introduced in the facility, in pursuit of finding optimal combination models using fuel cells, solar cells, and storage cells through experiments. These efforts, once spread among ordinary households, will effectively lead to reductions in primary energy consumption and waste heat to be released into the urban atmosphere.

To that end, the NEXT 21 facility will continue to serve as a testing site in quest of environmental conservation, energy saving, and affluent lifestyles.

3.1.2 Countermeasures against heat island phenomena at Namba Parks

3.1.2.1 Overview of Namba Parks

Namba Parks stands on the former site of the Osaka baseball stadium, which had been long loved by citizens before it was demolished in 1998. After going through the first phase of construction completed in October 2003, and the second in April 2007, the office and shopping complex now consists of a low-rise commercial building and a high-rise office building. On the rooftop of the 10-story shopping mall is a

FIGURE 4.24

Photograph (A) and thermograph (B) of a cool spot in the courtyard taken at around 3:00 p.m. on August 12.

(Photography: Professor Emeritus Akira HOYANO).

terraced garden shown in Fig. 4.26 with an area of 11,500 m^2, stretching up from the ground level, which provides an overwhelming open-air space at the site with a floor area ratio of 800% sandwiched between railroad tracks and an express way. The 5300-square-meter greenery area covers about half the rooftop garden, including flowerbeds with various kinds of flowers to be enjoyed all year round, allotment gardens where citizens can grow vegetables or others, currently used as a barbeque area, and forests simulating local natural vegetation. Thus, the garden is carefully designed to serve people and local ecosystem.

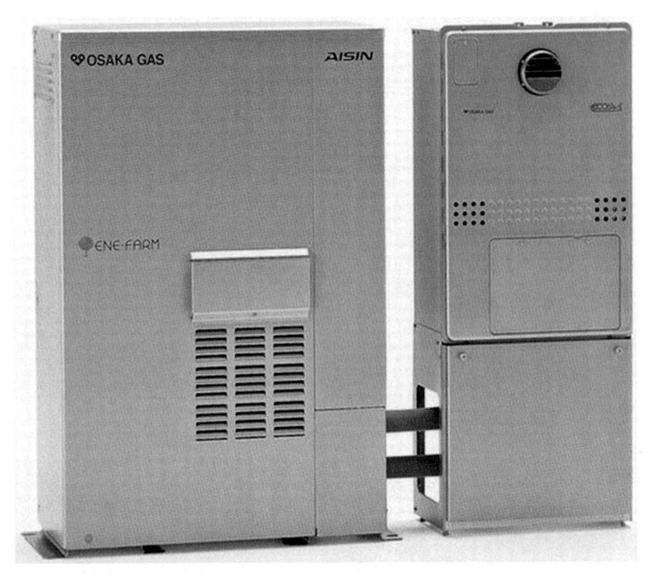

FIGURE 4.25

Solid oxide fuel cells—based cogeneration system.

FIGURE 4.26

Appearance of Namba Parks.

Since the so-called Minami area, centering around Namba, is said to have little greenery, the emergence of the vast green space, a precious source to combat warming, is expected to have a better microclimatic effect on the local environment. According to a multiple-year field investigation, the rooftop garden is found to have served as a pleasant place for local residents and visitors to comfortably rest and relax in various ways, where enough shade is secured through rich leafy shade and complicated wall arrangements in airy open sitting areas and pergolas, with strong winds being blocked.

3.1.2.2 Thermal comfort on the rooftop garden

Fig. 4.27 shows changes in the new standard effective temperature (SET*) people feel when moving along the walkways in the garden from the top floor downward [12]. Except at 11:30 soon after the facility opens when the walkways are exposed to the sunlight shining down right from the above, at other time settings people can enjoy strolling in a not-too-hot environment almost throughout the walkways, especially in the shady middle range. The walkways are much shaded with many tall trees planted around, and some are built through a forest.

Fig. 4.28 indicates thermal comfort measurements at many rest areas in the garden [13]. Both the densely and sparsely planted areas generally show lower feeling temperatures than open spaces, though they are switched around at some hours. In densely planted areas, despite low wind speed, the air is greatly cooled by the effect of evapotranspiration, lowering feeling temperatures as a whole.

3.1.2.3 Heat island mitigation due to man-made green areas

Fig. 4.29 shows a thermographic image taken at noon in August. As Namba Parks is surrounded by railroad tracks, blacktops, and an express way, you can see its great difference from them in surface temperature. The gap grew as many as 25°C between the planted areas on the green rooftop and the asphalt pavement. The planted surface temperature was lower than even the ambient temperature, proving that it can absorb heat from the city.

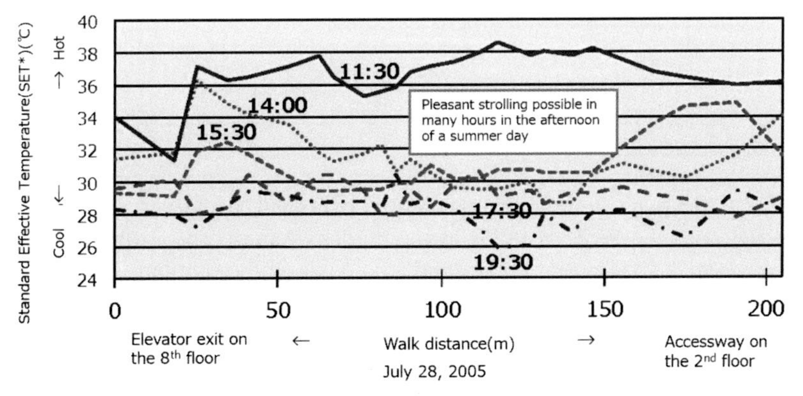

FIGURE 4.27

Thermal comfort at walkways in the garden.

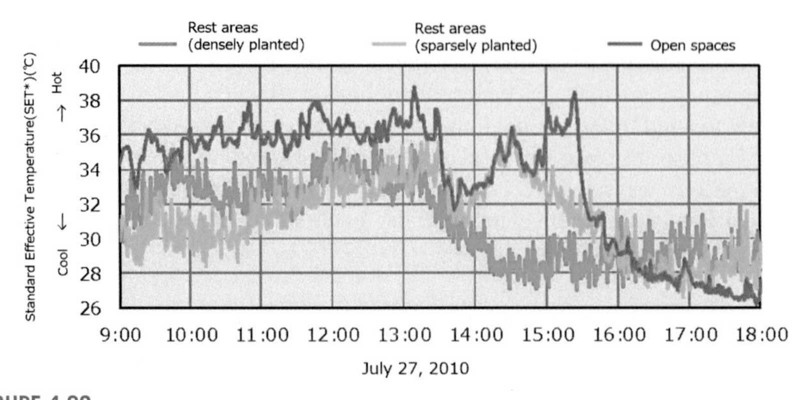

FIGURE 4.28

Thermal comfort at rest areas.

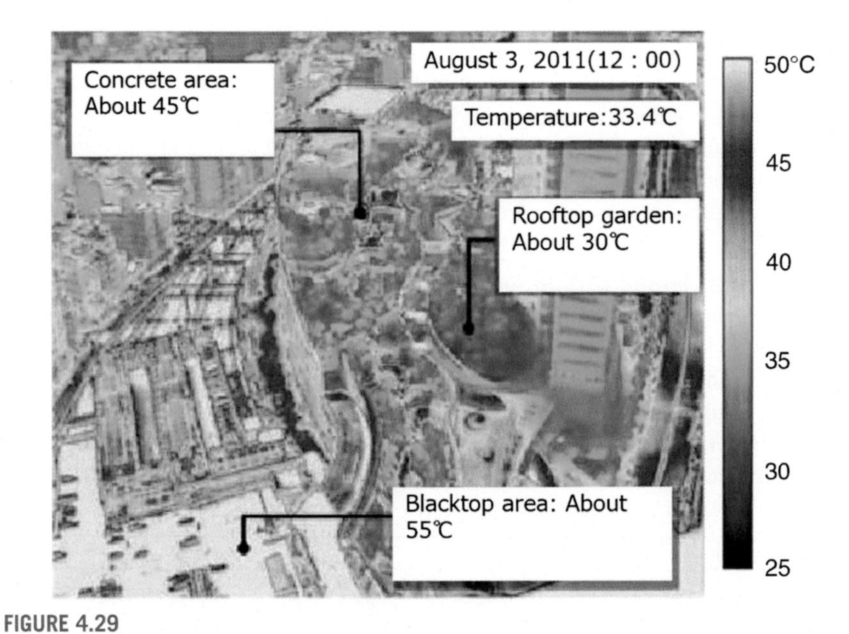

FIGURE 4.29

Thermography measurement at 12:00.

Fig. 4.30 shows line graphs indicting overnight decreases in temperature compared with the level at 18:00 in the garden. As opposed to the temperature drop of about 1.5°C on the rooftop until 6 o'clock in the morning, the forest area demonstrated the decline of about 3°C, suggesting a larger effect on temperature decrease. This reveals that large-scale green areas like Namba Parks can function as a cooling source in a city day and night.

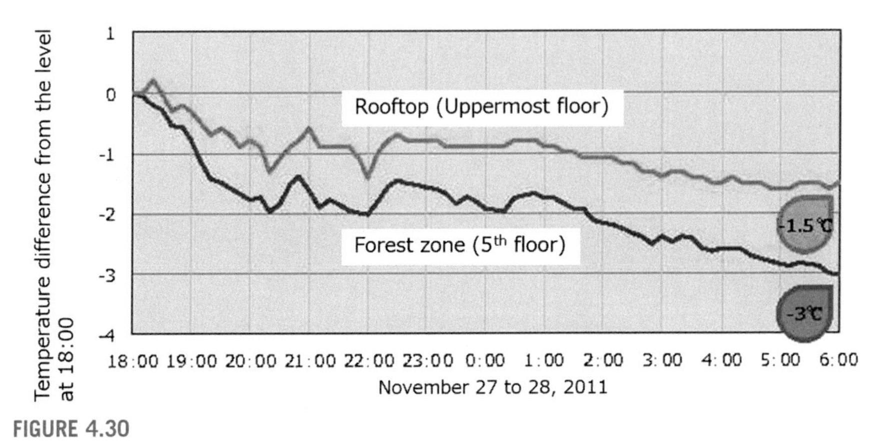

FIGURE 4.30

Nighttime temperature changes from 18:00 level.

In recent years, more cases have been seen in which people are more willingly to utilize their rooftops or artificial grounds as high-quality green areas. Changing such fixed concept of rooftops as being exposed to strong solar radiation and much heat, people tend to pay attention to them as a means of adding value to their assets in business development, which is expected to continue in the future. Also, amid the impacts of global warming and heat island phenomena becoming noticeable, setting green spaces bringing green shade on top of the buildings is highly effective in protecting the elderly, children, or other thermally vulnerable people from heat strokes. These data obtained from the research and study series are intended to provide valuable information to promote the development and spread of rooftop green gardens accessible to everyone in urban areas.

3.2 Ideas for the future development

A competition for an innovative idea on how to expand green shade and greening activities was held in 2009, in which opinions were widely sought from among the public. Among top entries, I would like to introduce the concept of "Edible Succession in Semba, incorporating Job, Food, and Home" proposed by the four-students team of Kobe University consisted of Ikuko Kunisue, Ryota Tanaka, Shigeki Tamochi, and Yuko Ishizu.

Focused on as its study site was Osaka's traditional downtown district called Semba, as shown in Fig. 4.31. As one of the core business centers in merchant city Osaka, the area enjoyed a long-time prosperity and served as home to a number of trading houses, securities firms, textile wholesalers, and banks. In recent years, however, quite a few office spaces have been left vacated due to changes in industrial structure. Such vacant or unused spaces in office buildings are largely distributed on the fifth or sixth floors with a height of about 15 m above ground or rooftop floors, in particular in Fig. 4.32. The winner concept suggests that such blank spaces, coupled with currently isolated green spaces sparsely spotted on some rooftops as shown in

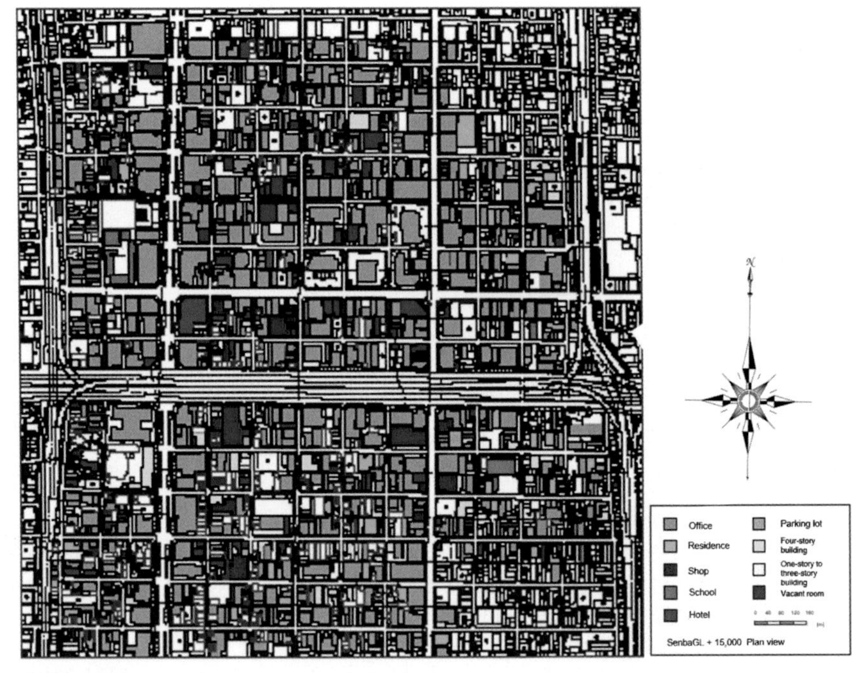

FIGURE 4.31

Location map.

Empty floor space in Semba

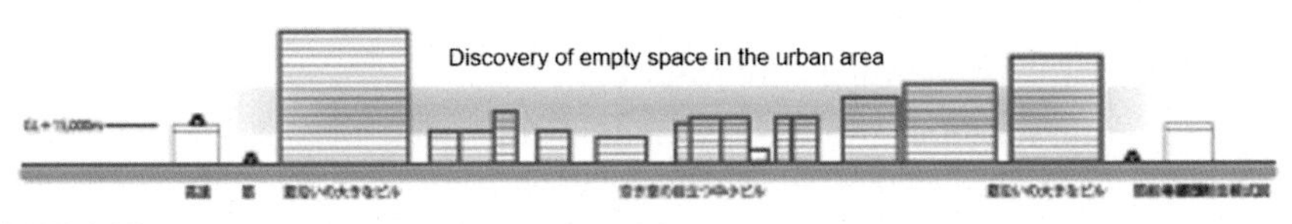

FIGURE 4.32

Empty space distribution.

Fig. 4.33, should contribute to creating cool and comfortable circuit-style walkways above ground, in a way that they can maintain two-dimensional and three-dimensional continuities in Fig. 4.34. The green passageways could also be ecologically utilized as migration channels for living things.

The process for designing the continuous green spaces in Fig. 4.35 should start with downsizing underused buildings, building semi-outdoor passages, and generating green walkways connecting to the rooftops of neighboring buildings. By repeatedly carrying out such procedures, a larger green refuge could be materialized, where office workers may be able to work outdoors, and have opportunities to communicate with

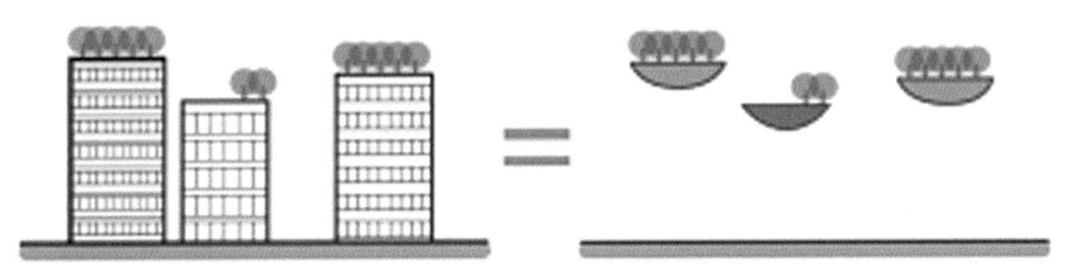

FIGURE 4.33

Isolated green space.

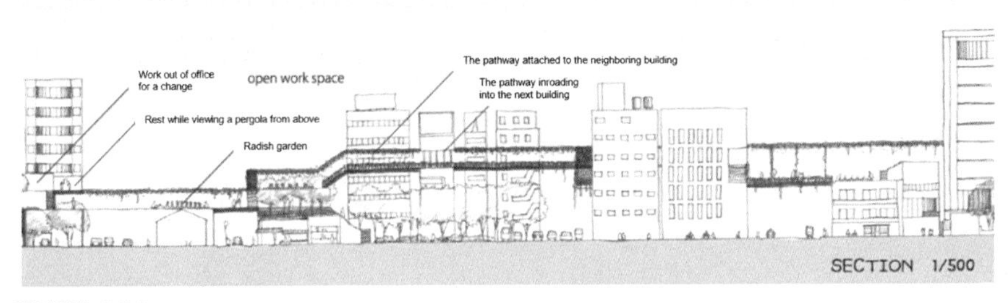

FIGURE 4.34

Three-dimensional development.

neighborhood residents engaged in greening activities (Fig. 4.36). Interactions between workers and residents could lead to community invigoration.

The proposal includes a program of growing herbs, vegetables, fruits, or other edible plants in built areas, in consideration of food problem, a key issue in the 21st century. The idea evolves around cultural aspects in agriculture, as well as the concept of symbiosis with ecosystem.

As shown in Fig. 4.37, the program offers you a closed-question flow chart which may help you find the most suitable activities for you among those involved in the green succession areas. Fig. 4.38 shows the completion image, in which the built vegetation helps mitigate heat effect, prevent cities from further falling into a mere mass of heat, and facilitate the revival of a city as working and living places filled with coolness, comfort, and richness (Tables 4.3 and 4.4).

4. Developmental idea using winds blowing along the river

4.1 Wind and temperature data based on an observation

Osaka City is located at the center of the Osaka Plain. This area has developed in the downstream region of the Yodo River flowing into Osaka Bay. Not only Yodo River but also large and small rivers flow in the center of Osaka city. A map of Osaka City and Osaka Bay (Fig. 4.39) shows the observation field [14] which is close to Namba and is surrounded by Dotonbori River and Kizu River leading to Osaka Bay located on the west side of the Osaka Plain.

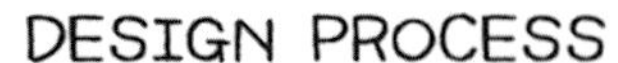

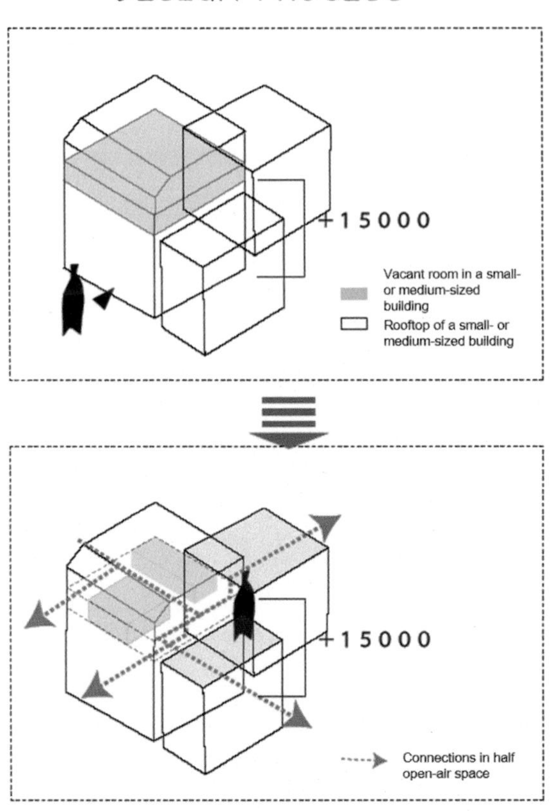

FIGURE 4.35

Design process.

For 1 month, August 2007, wind direction and speed were observed at the Point A and the Point B. Fig. 4.40 shows wind rose of each observation point on the ground and a near meteorological observation station on the top of the communication tower at 80 m above the ground. These wind roses indicate that the wind direction on breeze circulation days is between southwest to northwest where are the direction of Osaka Bay in most of the time regardless of the altitude. Fig. 4.41 shows a distribution of air temperature at 14:30 on 5 August 2008 and Fig. 4.42 shows a distribution of air temperature at 19:30 on 2 August 2008, which were interpolated based on the observation data [15]. At that time wind direction was between the northwest and southwest, and wind speed was 2–4 m/s during daytime and sunset. It was found that the temperature around Point A located downwind-side of Kizu River and area along Dotonbori River is relatively about 1 K lower than the surrounding area. It is important that citizens share this climatological information and use it for future city planning.

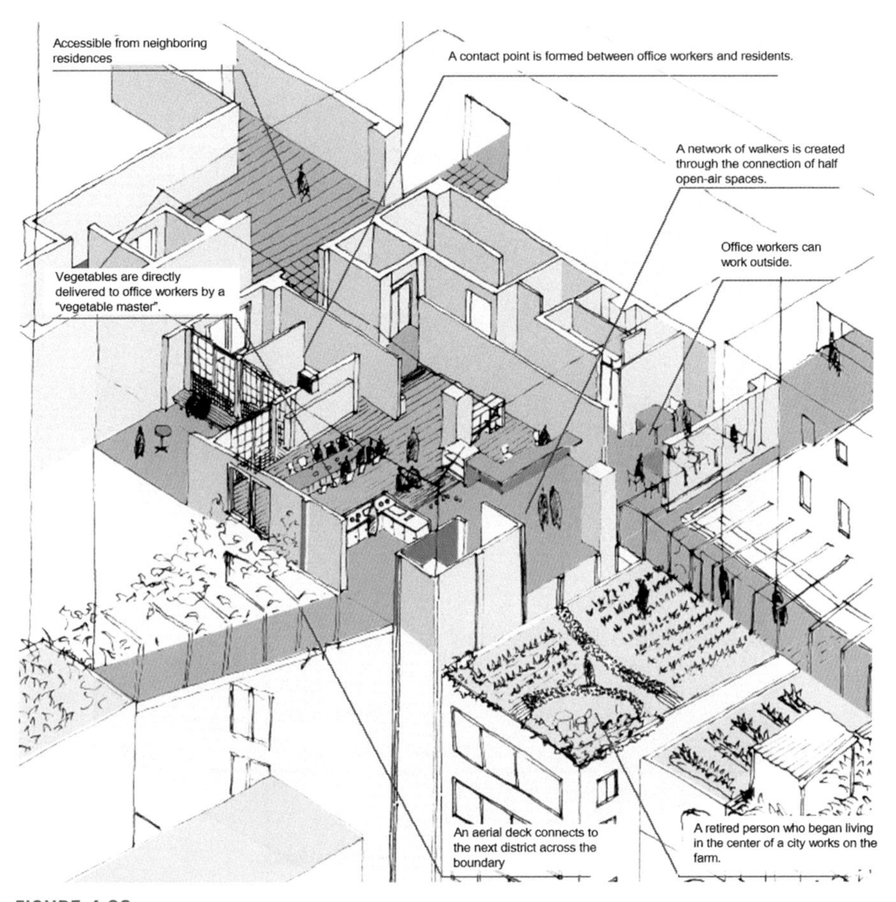

FIGURE 4.36

Succession.

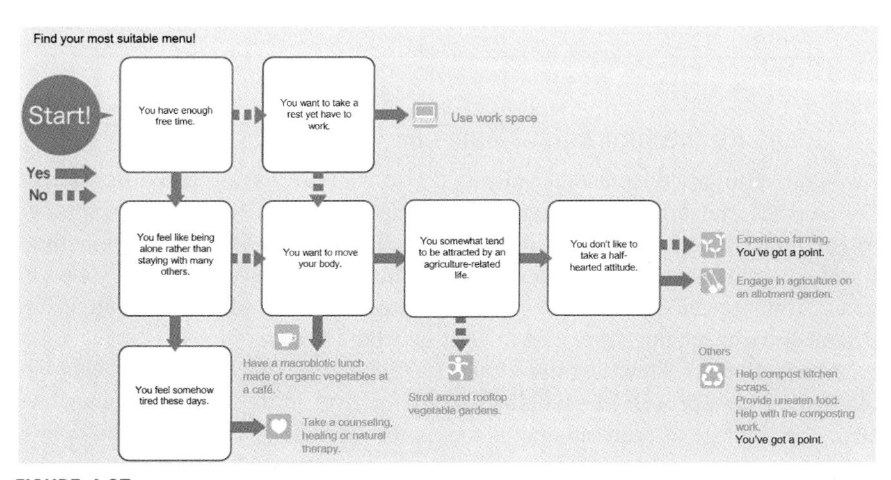

FIGURE 4.37

Program offering you a closed-question flow chart.

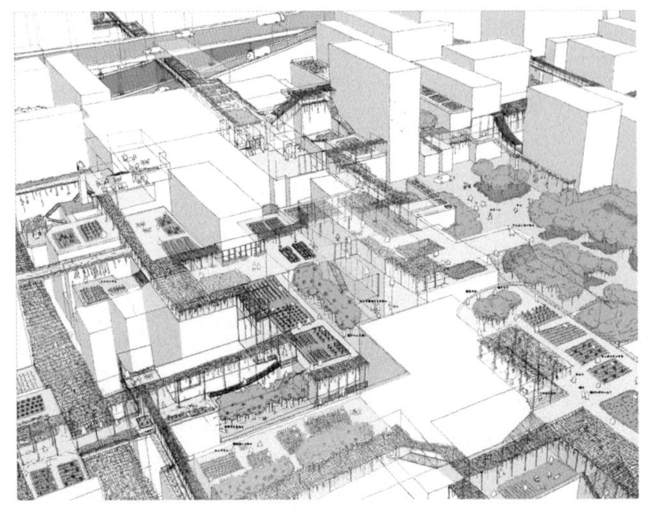

FIGURE 4.38

Completion image.

Table 4.3 Architectural overviewing of Next 21.

Site area: 1542.9m^2 Floor area: 4577.2m^2 Six stories above ground and one below	Building area: 896.2 m^2 Planted area: about 1000 m^2

Table 4.4 Architectural overviewing of Namba Parks.

Site area: 33,729 m^2 Rooftop area: 11,500 m^2 Ten stories above ground and four below	Building area: 25,500 m^2 Planted area: 5,300 m^2 100,000 plants of about 500 kinds

4.2 Ideas for the future development

I would like to introduce another idea that also won the 2009 competition, named "Wind paths stretching from water edges through neighbors" in Fig. 4.43 proposed by Koujirou Nakatsuji who was a graduate student of Kansai University at that time. It was a developmental idea using winds blowing along the river. Fig. 4.44 shows its focus strip, located between the Midosuji street and the Aiau bridge along the Dotonbori river running from east to west in central Osaka.

Although winds blow in from Osaka Bay along the Dotonbori River, they are blocked by rows of wall-like buildings along the river in the Dotonbori downtown area. The frontage of each building is too narrow to make a space for wind to pass

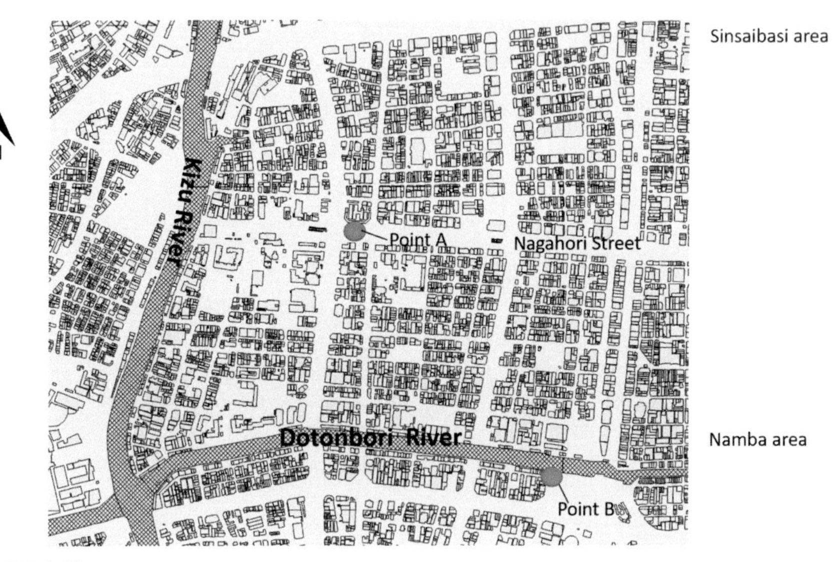

FIGURE 4.39

Map of the observation area.

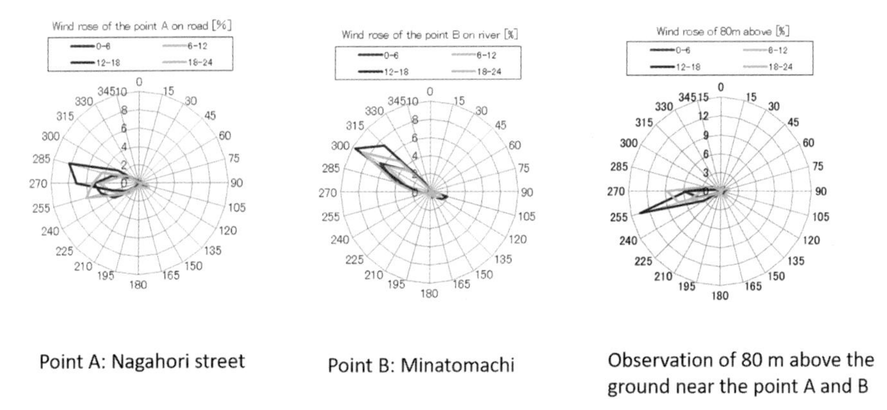

FIGURE 4.40

Wind rose based on the observation data in August 2007.

through. In these surroundings, the proposal shows the possibility of creating waterfront resort areas, where people can enjoy a cool breeze from the river. The areas forming wind paths can be realized only through cooperation between neighbors.

As illustrated in Fig. 4.45 and Fig. 4.46, "inverted V-shaped spaces" should be created two-dimensionally and three-dimensionally as well alongside of the river, through which river winds bring new ventilation in the neighborhood. The wind paths invite a flow of people and bring about water amenity as well as people's migration.

As shown in Fig. 4.46, these small inverted V-shaped spaces each created right above the boundary of two adjacent building sites provide an interface of a somewhat public nature between any flows of people or others.

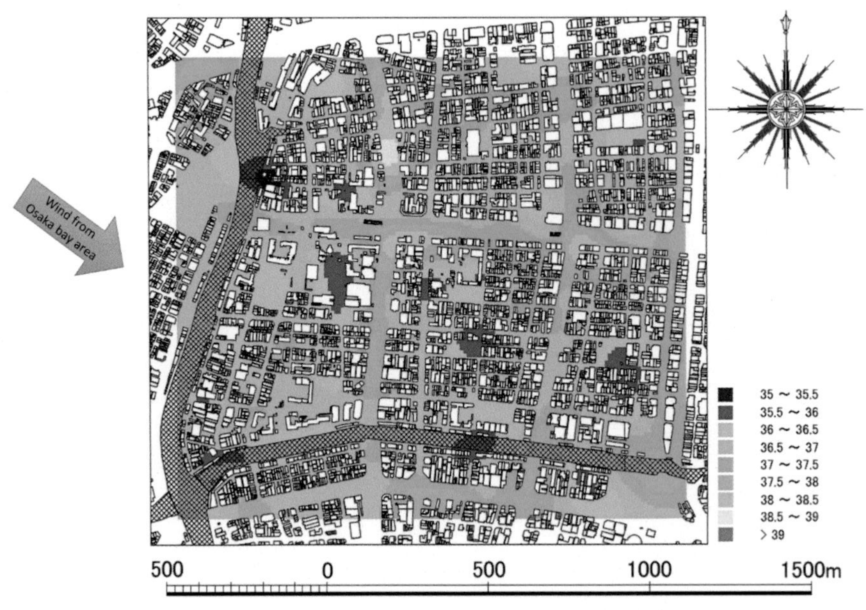

FIGURE 4.41

Temperature distribution based on the observation data at 14:30 on August 5, 2008.

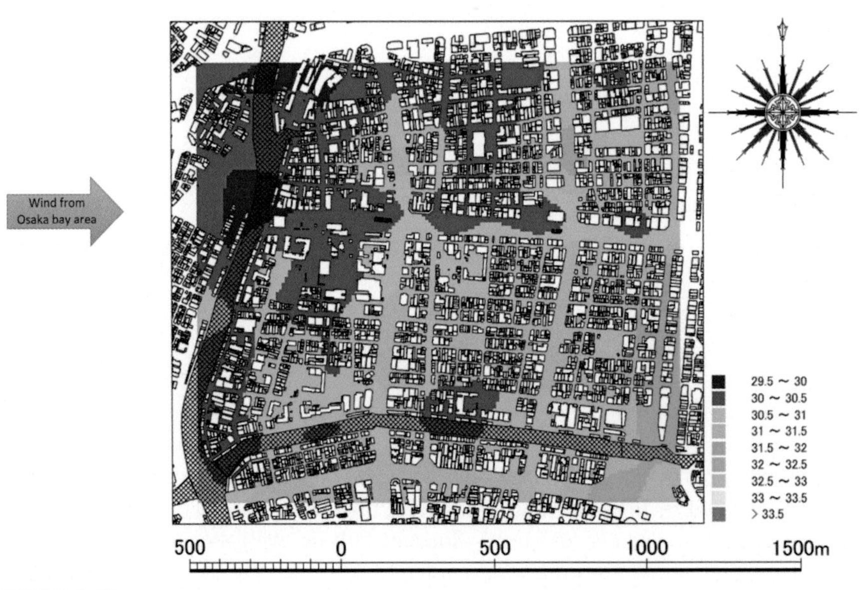

FIGURE 4.42

Temperature distribution based on the observation data at 19:30 on August 2, 2008.

FIGURE 4.43

Completion image.

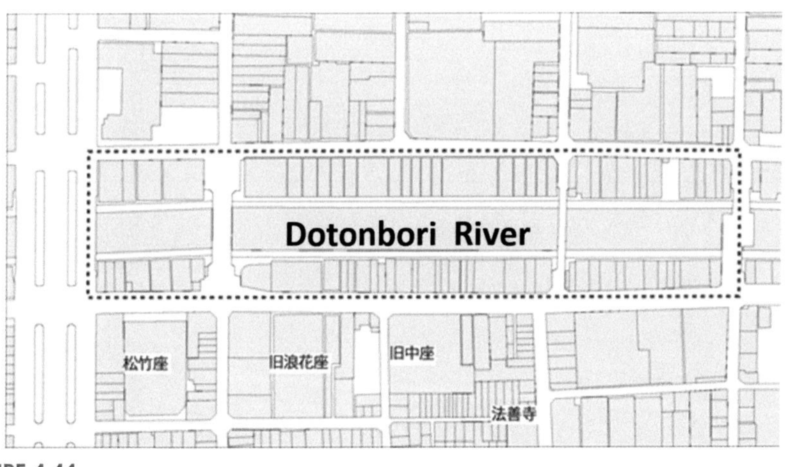

FIGURE 4.44

Location map.

Fig. 4.47 indicates the completion image and the completion model. As an example of cultural application, the small semipublic space may be possibly used as an event venue in collaboration with a nearby theater, as seen in Fig. 4.48. When applied as a business-related interface with the help of neighborhood commercial facilities, it may be turned into an open-air café, which is exemplified in Fig. 4.49.

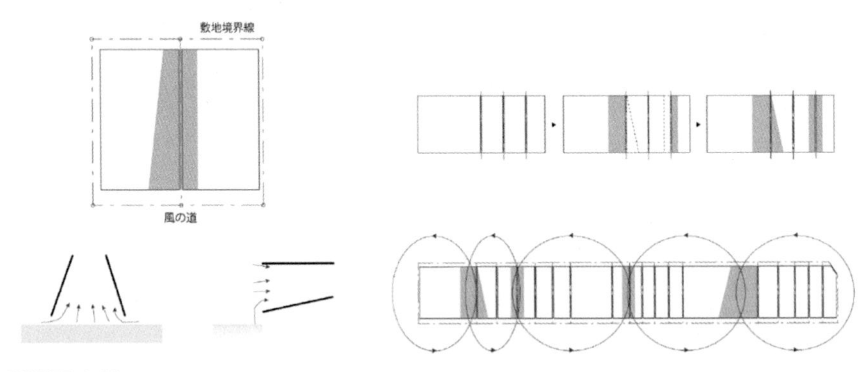

FIGURE 4.45

Basic concept "Inverted V-shaped space."

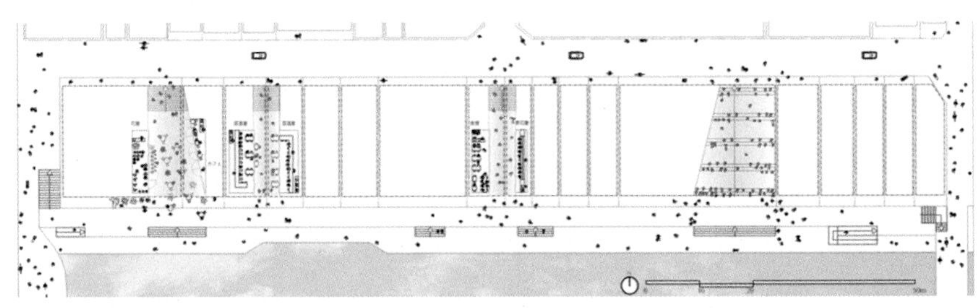

FIGURE 4.46

Plan view.

FIGURE 4.47

Completion image model.

FIGURE 4.48

Interface with culture.

FIGURE 4.49

Interface with business.

Thus, as indicated by Figs. 4.43–4.49, this proposal is designed to help revitalize the Dotonbori downtown area as a cool, comfortable and rich commercial district, by transforming over time the densely crowded riverside commercial spaces into porous or airy urban areas by means of the creation of inverted V-shaped wind paths.

References

[1] Osaka Prefecture Government. http://www.pref.osaka.lg.jp/chikyukankyo/jigyotoppage/heat_mati.html. Accessed August, 31, 2019.

[2] Osaka Prefecture Government. http://www.pref.osaka.lg.jp/kannosuisoken/hakusyo/osaka_kankyo_2004.html. Accessed August, 31, 2019.

[3] Google Map. https://www.google.com/permissions/geoguidelines/. Accessed August, 31, 2019.

[4] Osaka Prefecture Government.http://www.pref.osaka.lg.jp/kannosomu/midoritokazenogekan/coolspot.html. Accessed August, 31, 2019.

[5] Hoyano T. Protection against the heat and planting method. Heat and Environment. Tokyo: Dow Kakoh K.K.; 1988. p. 123–4.

[6] Nitta S, Azuma S, Ishii A. Design of micrometeorology in environment afforrestration. Tokyo: Kajima Institute Publishing; 1981. p. 40–2.

[7] Nishimura N, Nomura T, Iyota H, Kimoto S. Novel water facilities for creation of comfortable urban micrometeorology. Solar Energy 1998;64(4):197–207.

[8] Yoshinari K, Niwa H, Kunimatsu Y, Miura M. Study on the performance verification and evaluation of the district heating and cooling system using unutilized energy such as thermal energy of river water, report 1: the outline of the system and flamewowk of verification and evaluation. In: Proc. 2005 SHASE annual conference. SHASE; 2005. p. 2141–4 (in Japanese).

[9] Heat Pump and Thermal Storage Technology Center of Japan, Town creation by making use of "river water" Osaka Nakanoshima District – Creation of a low-carbon-emission town made possible by unused energy -, https://www.hptcj.or.jp/Portals/0/english/Learning/Nakanoshima%20DHC.pdf.

[10] Kaneko R, Hayashi H, Niwa H, Takahashi N, Koike K, Mishima N, Shimoda Y. Study on performance verification and evaluation of district heating and cooling system using thermal energy of river water, report 16: evaluation of exhaust heat load reduction causing heat island using thermal energy of river water and efforts of 11 years after starting operation. In: Proc. 2016 SHASE annual conference Kansai Branch. SHASE; 2016. A84 (in Japanese).

[11] "NEXT21" Edit committee, NEXT21: All about the NEXT21 Project, ISBN 4-7678-0491-4, Japan, 2005, pp.198-201.

[12] Akagawa H, Kubota T, Ichikawa K, Takebayashi H, Moriyama M. A large rooftop garden on a commercial building in Japan and its thermal environment during summer. In: Proceedings of the 2008 World Sustainable building conference, Melbourne; 2008.

[13] Akagawa H, Sugimoto H, Terai M, Makino M, Okazaki Y. Operation and management of a large rooftop garden on a commercial building in an urban area and multicomponent evaluation of environmental performance of the garden. In: Proceedings of the 2014 World Sustainable building conference, Barcelona; 2014.

[14] Nabeshima M, Nishioka M, Nakao M, Mizuno M, Tokura H, Mizutani S. Urban vegetation effects on the spatial variability of temperature in the city center. Web Journal of Heat Island Institute International 2012;7(2):126–33.

[15] Mizuno M, Nabeshima M, Nakao M, Nishioka M. Study on temperature distribution of an urbanized area using a vehicle with mobile measurement system: analysis on characteristics of the horizontal distribution using semivariogram and kriging. Journal of Environmental Engineering, AIJ 2009;644:1179–85 (in Japanese).

CHAPTER

Evaluation methods of adaptation cities

5

Shinji Yoshida, Ph.D[1], Atsumasa Yoshida, Dr.[2], Shinichi Kinoshita, Dr.[3]

[1]*Associate Professor, Department of Residential Architecture and Environmental Science, Nara Women's University, Nara, Japan;* [2]*Professor, Department of Mechanical Engineering, Osaka Prefecture University, Sakai, Japan;* [3]*Associate Professor, Department of Mechanical Engineering, Osaka Prefecture University, Sakai, Japan*

Chapter outline

Adaptation Measures for Urban Heat Islands. https://doi.org/10.1016/B978-0-12-817624-5.00005-1

1. Introduction

As mentioned previous chapters, we have outlined variable countermeasure techniques for adapting the negative effects of the heat island phenomena on the urban climate. We also introduced the applications of these countermeasure techniques to the practical urban and building planning. In this chapter, we outline methods to evaluate effects of installing the countermeasure techniques to these practical planning on the pedestrians in outdoor spaces.

1.1 Relationship between thermal environmental conditions and responses from pedestrians

For evaluation on effectiveness of adaptation techniques to the heat island phenomena, points of view for effects of these on the thermal comfort for pedestrians are required. In general, the strength of the heat island phenomena is defined on the differences of the air temperature between the urban and the rural areas. When we propose the adaptive techniques based on this definition, it is concerned that people attract only countermeasure techniques that have high potential to decrease the air temperature. However, the thermal comfort for the pedestrians is affected by various physical and mental elements including the air temperature.

Fig. 5.1 summarizes the relationship between the elements concerning the thermal environment and the responses by the pedestrians. In this figure, the elements and the conditions affecting the thermal and the comfort sensitivities are located at the upper side, while the physiological and the psychological responses at the

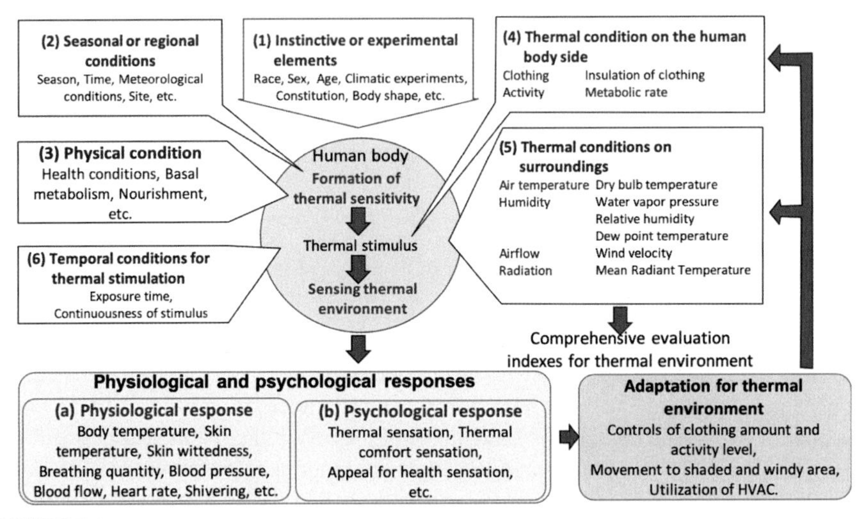

FIGURE 5.1

Relationship between thermal environmental conditions and responses from pedestrians.

lower side. The elements and the conditions are listed as follows: (1) instinctive and experimental elements, (2) background environmental conditions, (3) physical conditions, (4) thermal conditions for human body, (5) thermal conditions for surroundings, and (6) temporal conditions of thermal stimulus. In this figure, the air temperature that is primary focused on the countermeasure technique against the heat island phenomena is no more than one element in the (5) thermal conditions for surroundings. The elements of numbers (1), (2), and (3) also have large variations in the pedestrians. Hence, it is quite difficult to optimize the thermal and the comfort environment to each pedestrian. The designers and the planners of the adaptive cities are required both to recognize this difficulty and to have the point of views on regulating adequately the transitions of the elements concerning the thermal comfort, such as the air temperature, the wind velocity, the humidity, and the radiation.

2. Outline of mechanisms and modeling of human thermoregulation

The thermal comfort for the pedestrians has a strong relationship to the thermoregulation mechanisms in the human body. In this section, we outline both the basic theory and the modeling of the thermoregulation mechanisms in the human body.

2.1 Mechanisms of thermoregulation based on heat budget of human body

The thermoregulation mechanisms in the human body are significantly affected by the heat budget in it. This is due to the fact that humans are homeothermic animals and make sustained efforts to regulate the body temperature within the range approximately between 36°C and 37°C in order to remain in good health.

The equations for the heat budgets in the human body, the core layer, and the skin layer are respectively expressed as follows:

$$S = S_{cr} + S_{sk} = M + W + C + R + LE + Q_{RES} \tag{5.1}$$

$$S_{cr} = M + W - B - D + Q_{RES} \tag{5.2}$$

$$S_{sk} = C + R + E + B + D \tag{5.3}$$

where S, S_{cr}, and S_{sk} are the heat storages per the skin area in the human body, the core layer, and the skin layer (W/m^2), respectively. The values of these elements affect the increase or the decrease in the temperature of each element. In these equations, a positive value indicates heat storage of each layer, or an inflow of energy to the central part of the human body, except for the terms of the B and the D, as will be described later. The symbols M and W denote the amount of metabolic heat production (W/m^2) and that of mechanical work accomplished (W/m^2), respectively. These elements are primary included in the heat budget equation of the core layer (Eq. 5.2), due to being mainly generated in the central organs and the muscles. In general, the values of W show negative, but are negligible due to the significant small value than M. The symbol Q_{RES} denotes the amount of the heat exchange from respiration (W/m^2). Q_{RES} is the directional heat exchange between lungs in the central organs and the surrounding air. Thus, it is also included in Eq. (5.1). The symbols B and D are the amounts of the heat loss via the blood flow and the heat conduct from the core to the skin layers, respectively. For these elements, a positive value indicates the heat loss from the core layer to the skin one. These elements are eliminated from Eq. (5.1) by adding both sides of Eq. (5.3) to those of Eq. (5.2). The symbols C, R, and LE denote the convective, the radiant, and the evaporative heat exchanges between the skin layer and the surroundings (W/m^2).

The heat budget in the human body exposed to the cold or the hot environment affects the activation of the thermoregulation system in it. When a pedestrian stays in the cold environment without activating the thermoregulation, a large amount of heat is lost from the pedestrian's body to the surroundings. The sum of values between C, R, LE, and Q_{RES} becomes negative, and the absolute value of it surpasses that of the total amount of the heat generation in the body or the sum of both M and W. Hence, the value of S becomes negative, and the body temperature decreases. For reducing it, for example, the following two activations are occurred in the human body: the shivering of the muscles in the body and the vasoconstriction. The shivering generates the heat and contributes the increase of the amount of the metabolic heat M, while the vasoconstriction reduces the heat loss with blood flow from the core to the skin layers B. These activations cause the increase of the heat storage in the core layer S_{cr}.

On the other hand, when a pedestrian is exposed in the hot environment without activating the thermoregulation, the heat releases to the surroundings are reduced. The sum of the whole elements of the heat exchange between the human body and the surroundings reaches the positive value that indicates the inflow of the heat from the surroundings to the human body via the surface. This results in the increase of the body temperature because the value of S becomes positive. In this condition, the pedestrian feels the heat. For reducing it, for example, the following two activations occur in the human body; sweating and vasodilation. Sweating causes an increase of the absolute value of the amount of evaporative heat exchange LE when the water vapor pressure of the surrounding air is smaller than saturated water vapor pressure at the skin surface. Increase of the absolute value of LE causes the heat storage in the skin layer S_{sk} to decrease. Vasodilation also increases the heat loss with blood flow from the core to the skin layers B. The increase of B contributes the decrease of S_{cr}.

As mentioned above, the thermoregulation systems contribute that the value of S reaches nearly zero by regulating the values of both S_{cr} and S_{sk}. The activations of these systems cause the stress in the human body. Hence, the cold and the hot environment affect the thermal comfort for the pedestrians.

2.2 Human thermoregulation model

The human thermoregulation model enables us to estimate the various physiological mechanisms regulating the transients of the core temperature in the human body within the normal ranges. In this model, the physiological responses in the human body are calculated by using thermal environmental conditions around an entire body, such as air temperature, humidity, airflow, radiation, clothing, and metabolic rate. Recently, various types of the human thermoregulation model have been proposed by the researchers in the fields of medicine, civil engineering, and building environmental engineering.

The human thermoregulation models are classified by the following two types: the unit model for the entire body and the multinodes model with body parts. The model proposed by Fanger [1] and that by Gagge et al. [2] are well known as the unit thermoregulation model of the human body.

Fanger's model was developed for estimation of the thermal environment in the indoor space and has been used for the calculation of the predictive mean vote (PMV). Fanger's model has become to be well known with applications to the evaluation of the thermal environment using the PMV. However, the following two issues remain in this model when being applied to the evaluation of the thermal comfort in the outdoor spaces: (1) the accuracy of the model functions to the thermoregulation of the human body and (2) the method to the estimation of the core and the mean skin temperatures. With respect to the former one, the model functions to the thermoregulation in Fanger's model have high reliability because these functions were obtained by experimental results using enormous number of the subjects. However, the thermal conditions of these experiments primarily consider the thermal conditions in indoor spaces and not those in outdoor spaces, or the hot or the cold conditions in the summer or the winter seasons. With respect to the latter one, the mean skin temperature in Fanger's model is also calculated by using the function of the heat production from the metabolism. Hence, it is understood that the prediction accuracy of the mean skin temperature in Fanger's model is insufficient because the temperature does not include the effects of the heat budget in the human body. Consequently, it has been generally considered that the evaluation using the PMV is not adequate when the absolute value of it is larger than two.

On the other hand, the entire body in Gagge's model consists of a core layer and a skin layer. Hence, this model has been well known as the two-node model. The heat budgets of the body, the core, and the skin layers are evaluated by using Eqs. (5.1)−(5.3) in Section 2.2 in this chapter. The two-node model allows us to evaluate wide ranges of the thermal environment from the cold to the hot conditions because this model considers the following thermoregulation mechanisms: the heat release by sweating, the heat exchanges between the core and the skin layers by the blood flow, and the heat product based on the shivering. Hence, several researchers have proposed revised models based on the two-node model for the evaluations of the thermal environment in the outdoor spaces.

The multinode human thermoregulation model enables us to evaluate effects of the inhomogeneity of the thermal environment on the comfort for the pedestrians by calculating the distributions of physiological responses on segments of the entire human body. Stolwijk's multinode model [3] is one of the best known multinode human thermoregulation models. This model consists of six body segments divided into the following four layers: core, muscle, fat, and skin layers. Subsequently, various other advanced models have also been developed by reference of it (Gordon et al. [4], Smith [5], Fu [6], Huizenga et al. [7]). Recently, a research group led by Tanabe in Waseda University in Japan also developed four multinode human thermoregulation models known as "65MN" (Tanabe et al. [8]), "COM" (Tanabe et al. [9]), "JOS" (Sato et al. [10]), and "JOS-2" (Kobayashi and Tanabe [11]).

The application of the multinode human thermoregulation model has the following two advantages: (1) the availability to evaluations of the inhomogeneous thermal environment and (2) the improvement of the prediction accuracy of the thermoregulation by blood flow. According to the former one, for example, the entire body of the JOS-2 model [11] is divided into 17 body segments (head, neck, chest, back, pelvis, left shoulder, left arm, left hand, right shoulder, right arm, right hand, left thigh, left leg, left foot, right thigh, right leg, and right foot). All individual body segments consist of a core layer, a skin layer, an artery blood pool, a vein blood pool, and a superficial vein blood pool. The head segment additionally includes a first and a second layer in order to consider effects of rapid physiological response to changes in the thermal environment. On the other hand, according to the latter one, the hands and the feet segments also include arteriovenous anastomoses (AVAs) connecting arteries and superficial veins without exchanging heat with any tissues. The AVA contributes to the rapid heat release from four limbs by dilating when the pedestrian is exposed to the hot environment. It is hoped that these functions improve the prediction accuracy of the evaluation of the thermal environment in the outdoor spaces in the summer season. Hence, it is expected that the multinode human thermoregulation model will be applied to the evaluation of the thermal environment in the adaptive cities, although we only show application examples of the two-node model introduced above as the unit model.

3. Evaluation scales of adaptation cities

At the previous sections, we described the necessity of considering the physiological responses to the thermal environment for evaluating the effectiveness of the countermeasures and the planning on the adaptive cities and buildings. The methods of modeling the human thermoregulation mechanisms were also outlined there. In the present section, we introduce the following several indexes for evaluations of the thermal environment in the adaptation cities: (1) wet-bulb globe temperature (WBGT), (2) new effective temperature (ET*), and (3) new standard effective temperature (SET*).

3.1 Wet-bulb globe temperature

The WBGT is a well-known evaluation scale for prevention against hyperthermia. This was established for evaluations of hyperthermia in US armies and was based on the functions from the experiments under the severe hot environment. The WBGT is defined as follows for the outdoor and for the indoor environment, respectively:

$$WBGT = \begin{cases} 0.7T_W + 0.2T_G + 0.1T_D & (\textit{for outdoor environment}) \\ 0.7T_W + 0.3T_G & (\textit{for indoor environment}) \end{cases} \tag{5.4}$$

where T_D is the dry-bulb temperature (°C), T_W is the naturally ventilated wet-bulb temperature (°C), and T_G is the globe temperature (°C).

According to evaluation scales, the WBGT remains the following two issues: (1) the method to measure the naturally ventilated wet-bulb temperature and (2) the reasonability to the human thermoregulation mechanisms. The naturally ventilated wet-bulb temperature is originally measured with the naturally ventilated wet-bulb thermometer that is left exposed to the solar and the long-wave radiations. However, the automatic continuous measurement of it is quite difficult. Hence the WBGT is often calculated by using the thermodynamic wet-bulb temperature as the T_W. It is obviously necessary to pay attention that the value of the thermodynamic wet-bulb temperature is different from the natural ventilated one. On the other hand, the equations for the WBGT are not based on the heat budget in the human body because the functions for it were derived from the measurement results. However, it has been also noted that the functions of the WBGT enables us to well express the conditions of hyperthermia for the pedestrians exposed to the hot environment. Hence, it is necessary to take care of the applications of the WBGT to evaluations on the normal thermal environment because the WBGT was originally proposed for the needs to evaluate the scorching hot environment.

3.2 New effective temperature and new standard effective temperature

The new effective temperature (ET*) and the new standard effective temperature (SET*) are evaluation scales of the thermal comfort and were proposed by

Gagge et al. [2]. These scales are defined as the dry-bulb temperature of the standard thermal environment at 50% RH where a combination of thermal sensation and total heat flux at the skin surface of an entire body equals that in the actual thermal environment. In these scales, it is considered that both the thermal sensation and the thermal comfort for pedestrians are determined by the combination of mean skin temperature (θ_{sk}), the mean skin wettedness (w_{sk}), and the total heat flux at the skin surface (Q_{sk}). Based on this consideration, the Q_{sk} in the actual thermal environment is defined as follows:

$$Q_{sk} = -\{(R+C)+LE\} = -\{\alpha_t f_{cl} F_{cl}(OT-\theta_{sk}) + \alpha_e f_{cl} F_{pcl} w_{sk}(pa - pSat_{sk})\} \tag{5.5}$$

where C, R, and LE are the convective, the radiant, and the evaporative heat exchanges between the skin layer and the surroundings (W/m^2), as mentioned in Section 2.1 in this chapter. The symbols α_t also denotes the sensible heat transfer coefficients at the skin surface (W/[m^{2}K]), while α_e the latent heat transfer coefficients at the skin surface (W/[m^{2}kPa]). Furthermore, OT is the operative temperature (°C); pa is the vapor pressure in ambient air (kPa); $pSat_{sk}$ is the saturated water vapor pressure at the skin surface (kPa); f_{cl} is the clothing area factor; F_{cl} is the intrinsic clothing thermal efficiency; and F_{pcl} is the permeation efficiency.

Under the consideration mentioned above, we can find lots of variations of the thermal environmental conditions determined to be equivalent in the thermal sensation. In calculations of ET* and SET*, the operative temperatures are first calculated by using both the thermal sensations from the results of the calculations in the actual environment and the thermal environmental conditions standardized by altering several elements of the conditions. The scales of ET* and SET* are estimated by exchanging these operative temperature to the equivalent temperature for the thermal sensation. The thermal environmental condition consists of the following elements as described in Fig. 5.1: (1) thermal conditions on the human body side (clothing amount, and metabolic rate) and (2) thermal conditions on surroundings (air temperature, relative humidity [RH], wind velocity, and mean radiant temperature). In calculation of the ET*, the value of the RH in the actual environment is only standardized to 50%. On the other hand, in that of the SET*, the values of the RH, the wind velocity, the metabolic rate, and the clothing amount are standardized to 50%, 0.10 m/s, 1.0 met, and 0.6 clo, respectively. The standard environmental conditions are supposed to the typical thermal conditions in the office rooms. Based on the consideration mentioned above, the Q_{sk} for the calculations of ET* and SET* are defined as follows:

$$Q_{sk} = -\{\alpha_t f_{cl} F_{cl}(ET^* - \theta_{sk}) + \alpha_e f_{cl} F_{pcl} w_{sk}(0.5pSat_{ET^*} - pSat_{sk})\} \tag{5.6}$$

$$Q_{sk} = -\{\alpha_{tS} f_{clS} F_{clS}(SET^* - \theta_{sk}) + \alpha_{eS} f_{clS} F_{pclS} w_{sk}(0.5pSat_{SET^*} - pSat_{sk})\} \tag{5.7}$$

where subscript S indicates the standardized variable.

In the comparison of the general versatility between ET* and SET*, it has been considered that SET* is more useful than ET*. The standard thermal environment in the calculation using ET* is affected by the wind velocity, the metabolic rate, and the clothing amount in the actual thermal environment, while that in the calculation using SET* is not affected. Hence, SET* enables us to carry out more standardized evaluation of both the thermal environment and the thermal comfort for pedestrians.

An issue of applications of SET* is the validation of the evaluation on the extremely hot environment where lots of pedestrians are exposed to the higher risk for the heatstroke. However, at the present, we cannot find the reasonable evaluation scales on the risk of the heatstroke based on the heat budget in the human body. At least, it is considered that the SET* is more effective scale for the evaluation on the hot environment than the PMV because the two-node model enable us to more accurately predict the mechanisms of the thermoregulation in the hot environment than Fanger's model. Hence, at the actual condition, we involuntarily apply the SET* to the evaluation on the extremely hot environment. In general, the two-node model is applied to calculations of the mean skin temperature (θ_{sk}), the mean skin wettedness (w_{sk}), and the total heat flux at the skin surface (Q_{sk}). However, it has been pointed out that the accuracy of the two-node model falls when the pedestrians act with the high metabolic rate with the secretion of the sweat. Hence, it is considered that the additional modifications of it or the applications of other thermoregulation models are required [12].

3.3 Remarks of application of evaluation scales to thermal environment in adaptation cities

In the present section, the several scales for evaluations of the thermal environment in the adaptation cities are outlined. Table 5.1 summarizes the comparisons between the WBGT and the SET* for evaluation of countermeasure method to adaptive cities against climatic changes.

4. An evaluation method based on field measurement

4.1 Measurement at Midosuji in Osaka city [12]

A fixed-point measurement on the thermal environment was conducted on the west sidewalk of Midosuji block in Chuo-ku, Osaka, as shown in Fig. 5.2. The heat balance of the street surface, the radiation environment of the street space, and the tendency of wind speed and direction were evaluated. Thermal comfort was evaluated by subjective reports and subject experiments. The thermal sensation received from the urban block was also evaluated based on the results of fixed-point measurement. The sidewalk is a tiled pavement with a width of about 6 m. Ginkgo biloba trees are planted in two rows on one side as street trees along the side road. The measurement date and time are two periods from 15:00 on August 9 to 18:00 on 10th and from 13:00 on August 30 to 18:00 on 31st. Fig. 5.3 shows the arrangement of the measuring device, and Fig. 5.4 shows the measurement status.

Table 5.1 Comparisons between indexes for evaluation of countermeasure method to adaptive cities against climatic changes.

	WBGT	SET*
Main target for evaluation	Extremely hot environment	From cold environment to hot environment
Input data	Dry-bulb temperature, naturally ventilated wet-bulb temperature, globe temperature	Air temperature, relative humidity (RH), wind velocity, mean radiant temperature (MRT), clothing amount, metabolic rate
Application of thermoregulation model	The application of the thermoregulation model is not required	The two-node model proposed by Gagge et al. [2] is usually applied to the calculation. The other models such as the multinode human thermoregulation models also allow us to apply it
Consideration of the heat budget in the human body	The WBGT does not consider the heat budget in the human body	The standard effective temperature (SET*) is calculated from the results of the thermoregulation models considering the heat budget in the human body
Validity of the applications to the evaluations of the extremely hot environment	The WBGT enables us to well express the conditions of hyperthermia for the pedestrians staying on the hot environment, because the WBGT was originally proposed to evaluate the scorching hot environment, like the environment for the blue collar workers	The SET* has not been validated for the evaluation on the extremely hot environment
Issues for evaluations on adaptation techniques to the hot environment in the urban area	**(1)** The WBGT does not enable us to evaluate contributions of each thermal environmental element to the thermal comfort for the pedestrians because it is not based on the heat budget in the human body **(2)** It is difficult to continuously measure the naturally ventilated wet-bulb temperature	The validity of the applications of the SET* to the extremely hot environment has not been judged

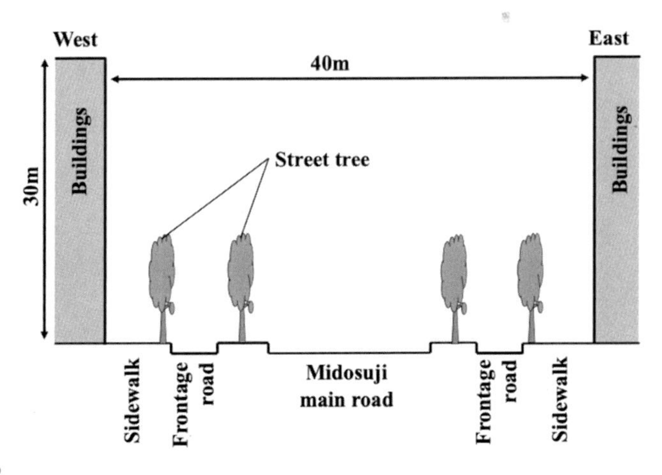

FIGURE 5.2
Outline of Midosuji road.

WBGT (wet-bulb globe temperature) index [13], one of the evaluations of thermal stress of workers in hot environments, is calculated from the results of fixed-point measurements (wet-bulb temperature, globe temperature, air temperature) during the first measurement period. Fig. 5.5 shows the change over time. The graph was displayed as a 10 min moving average. At the same time as the fixed-point measurement, the heat stroke index meter was measured every 3 h except for midnight,

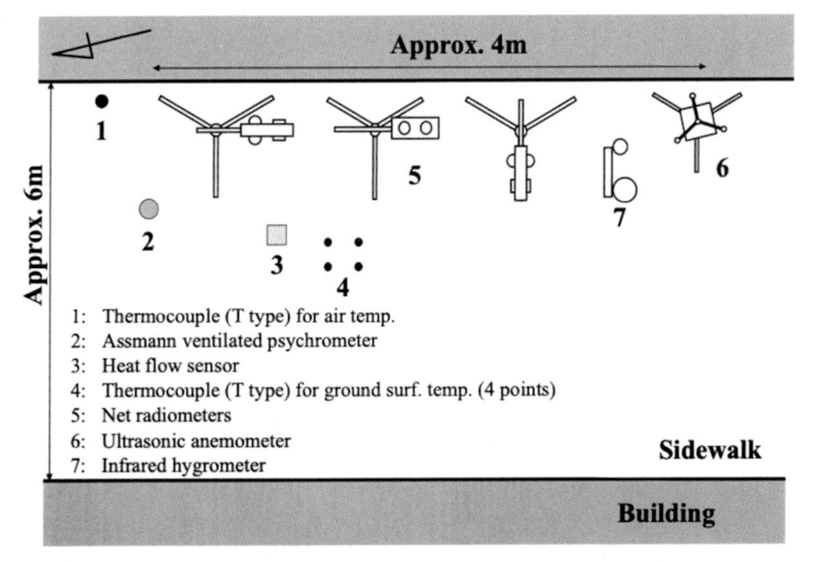

FIGURE 5.3
Measurement location and arrangement of measurement equipments.

FIGURE 5.4

Photos of measurement location.

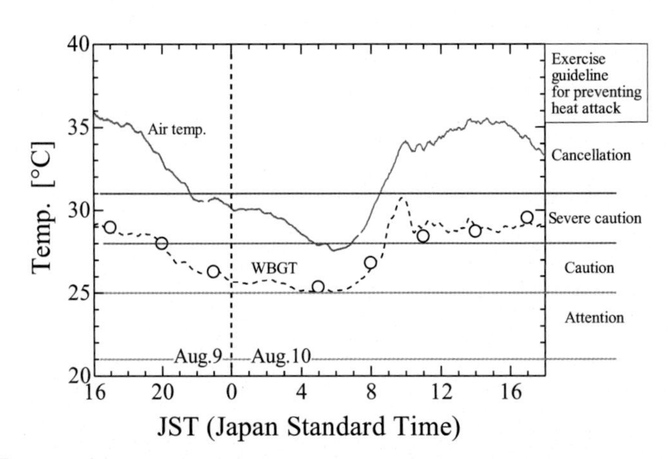

FIGURE 5.5

Change in wet-bulb globe temperature (WBGT) index over time and exercise guideline.

and there was good agreement. In addition, exercise guidelines for preventing heat stroke are also displayed on the right side of Fig. 5.5. From the start of measurement on August 9 to around 7 p.m. and from 9 a.m. on August 10 to the end of the measurement, it can be seen that the situation was "strict alert" (28–31°C). At the same time, a subjective report (street questionnaire) survey was conducted around the measured place. The implementation time is about 2 h from 1 p.m. on August 10 and about 3 h from 2 p.m. on August 31. The number of questionnaires collected is about 20 people. The results of thermal sensation and comfort are shown in Fig. 5.6. As can be seen from this figure, the 10th was hotter and more unpleasant. On August 31, 2006, the average skin temperature was evaluated by a subject experiment on a sidewalk where fixed-point measurement was performed. The test subject was a male in his 20 s and was clothed in short-sleeved shirts, long trousers, and sandals. Skin temperature was measured in a standing position, and a T-type thermocouple was used to measure seven points on the forehead, upper arm, back, abdomen, thigh, lower leg, and instep for 30 min at 1-min intervals. The average skin temperature was calculated by weighting these values based on the Hardy–Dubois 7-point method [14]. Moreover, the average skin temperature was predicted

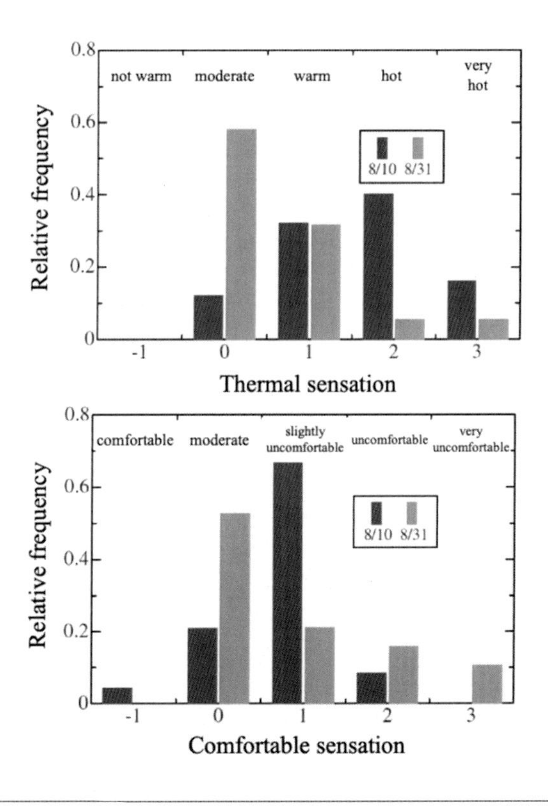

FIGURE 5.6

Questionnaire results for thermal and comfortable sensations.

based on the human body model (65-part body temperature regulation model [8]) from the fixed-point measurement results of meteorological elements (air temperature, humidity, wind speed, solar radiation) at the same point. The effect of solar radiation was evaluated based on literature [15]. Fig. 5.7 shows the results of average skin temperature measurement and model calculation. The result included data for 30 min from 1 p.m. The measurement results on August 31 and the calculation results were in good agreement. The average skin temperature on August 10 was predicted using the same model, and the result was higher than that on the 31st. It agrees with the result of the subjective report for passers-by.

4.2 Evaluation of thermal sensation in several outdoor radiation environments [16]

In this section, the thermal sensations of stationary human test subjects remaining under stable test environments for a relatively long period of time were evaluated. The thermal environments of the test locations were evaluated using the human thermal load, which can be used to quantitatively evaluate the heat transfer between the human body and the surrounding environment, and link the results to the thermal sensation. Among the mitigating or adaptive measures against the heat island phenomenon currently under consideration, to clarify the thermal stress on the human body in different radiation environments and the influence on the thermal sensation, human subject experiments were conducted at locations simulating the ground surface materials with different amounts of solar reflectance and with the formation of a tree canopy by tall trees, all of which affect the radiation environment.

To evaluate the thermal environment of the human body, the environmental factors and physiological state must be evaluated. This study built a human thermal sensation index assuming that the heat balance of the human body and the surrounding environment [15] are relevant to the thermal sensation. The index assumes that

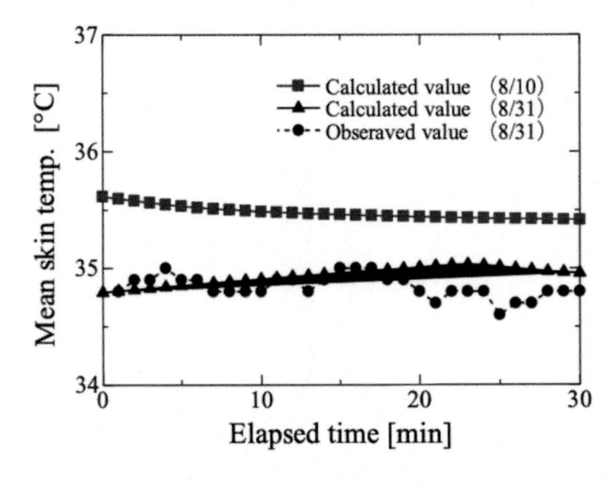

FIGURE 5.7

Comparison of measured and calculated values of mean skin temperature using human body thermal model.

the human heat balance consists of five factors: the metabolic rate, M; mechanical work rate, W; net radiation, R_{net}; sensible heat exchanges between the skin layer and the surroundings, C; and latent heat exchange between the skin layer and the surroundings, E, and the following equation is true under a heat equilibrium state.

$$M + W = -R_{net} - LE - C \tag{5.8}$$

In a nonequilibrium state, the residual of the five heat balances is given as the thermal load, and the state in which heat transfers to the human body is defined as a positive human thermal load.

$$F_{load} = M + W + R_{net} + LE + C \tag{5.9}$$

The human subject experiments took place outdoors and in a climate chamber. The outdoor measurements were conducted at the Nakamozu Campus of Osaka Prefecture University and the climate chamber measurements in the facility at the Izumi branch of the Osaka Research Institute of Industrial Science and Technology.

The experiments were conducted simultaneously at neighboring locations with different ground solar reflectance. The ground surfaces were a wood deck (solar reflectance $r = 0.20$) and tiles ($r = 0.31$) during the summer experiments. The ground surfaces were covered with a waterproof sheet ($r = 0.10$) and a white sheet (2.8 m long × 3.6 m wide) ($r = 0.76$) placed on the waterproof sheet during the winter experiments. The geometric factor from each subject's viewpoint was $f = 0.73$. The experimental locations were sufficiently spacious and open. The experiments were conducted for 2 days each during the summer and winter, namely, on July 28 and 29 and on November 30 and December 2, respectively. There were six test subjects, and the experiments took place between 11:00 and 14:00. With two subjects combined as a pair, three pairs were tested per day. The subjects switched locations on the second day for both the summer and winter experiments. The weather on the experiment days was calm and clear during both the summer and winter. The air temperature and humidity were almost the same during both days of the summer and winter experiments.

The ground surface inside the climate chamber was concrete, and thus, white and black sheets were placed in addition to the bare surface to test three different conditions of solar reflectance. The sheets were 2.8 m long and 3.6 m wide. The environmental factors of the climate chamber were as follows: air temperature of 30°C, relative humidity of 50%, wind speed of 0.3 m/s, irradiation from artificial lighting consisting of 67 metal halide lamps (2.0 × 2.7 m), solar radiation of 880 W/m^2, and infrared radiation of 587 W/m^2. The solar reflectance of the bare concrete floor, and the white and black sheets, measured at 1.0 m above the ground was $r = 0.10$, 0.33, and 0.02, respectively, including that of the surrounding area, which was not irradiated by artificial lighting. The subjects stood beneath the center of the artificial lighting. Four subjects participated in the experiments.

Fig. 5.8 shows the differential in human thermal load according to the respective ground surfaces during the summer and winter experiments. During the summer experiments, the mean differential of the human thermal load was 29 W/m^2 between

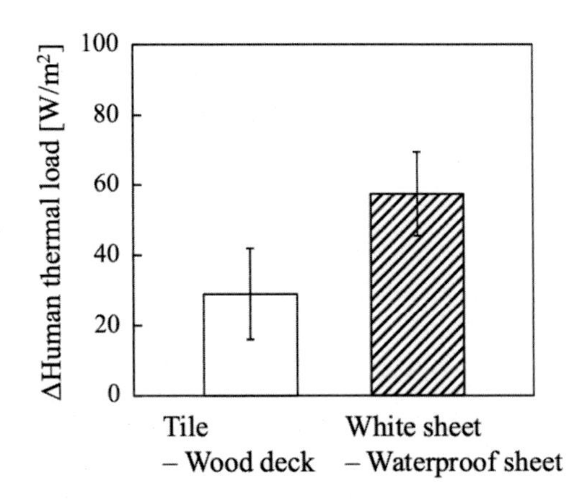

FIGURE 5.8

Human thermal load differential for various ground solar reflectance in outdoor.

the wood deck ($r = 0.20$) and tiles ($r = 0.31$), and the mean differential of the winter experiments was 57 W/m^2 between the waterproof sheet ($r = 0.10$) and the white sheet ($r = 0.76$). This indicates that the greater the solar reflectance, the greater the differential in human thermal load that occurs. Fig. 5.9 shows the thermal sensation on the respective ground surfaces during the summer and winter. The thermal sensation values reported on the ground surfaces with higher solar reflectance were higher. This also indicates that the differential in the thermal sensation is greater in the winter than in the summer, when the differential in the human thermal load is greater.

Fig. 5.10 shows the human thermal load according to the respective ground surfaces. The mean human thermal loads on the respective ground surfaces were greater in descending order of the solar reflectance (white sheet → concrete → black sheet). A t-test ($P < .01$) conducted on the human thermal load according to the

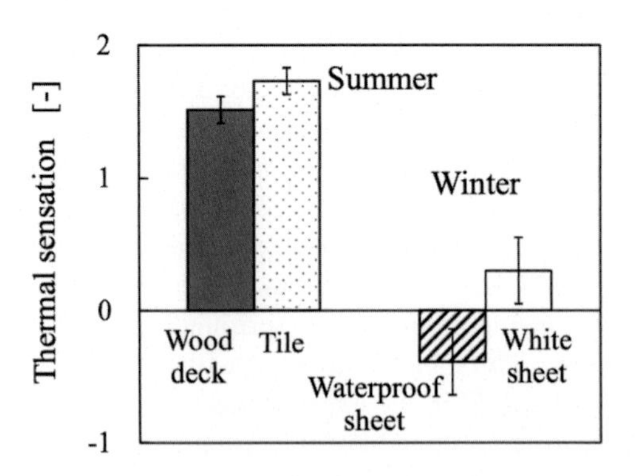

FIGURE 5.9

Thermal sensation for various ground solar reflectance in outdoor.

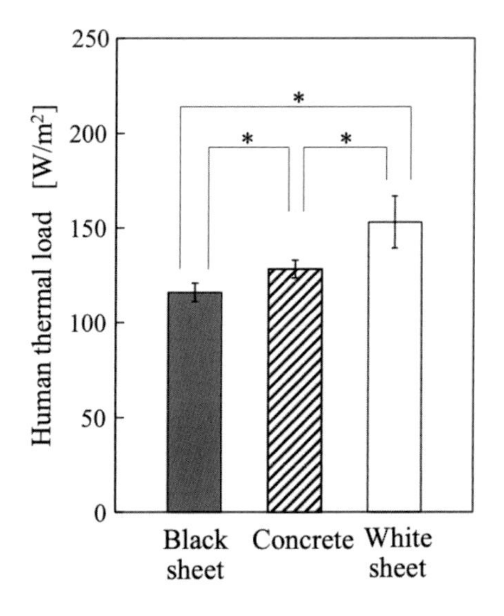

FIGURE 5.10

Human thermal load for various ground solar reflectance in climate chamber. The asterisk * means t-test ($P < 0.01$).

respective ground surfaces confirmed a significant difference. The results of the thermal sensation reported by the test subjects are shown in Fig. 5.11. A t-test ($P < .05$) was also conducted on the reported thermal sensation values for the respective

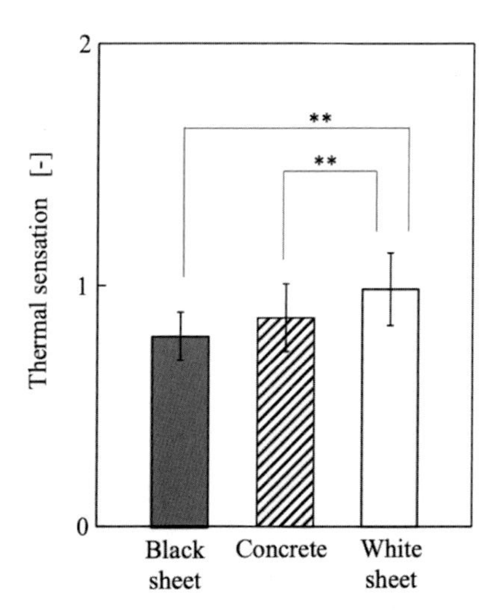

FIGURE 5.11

Thermal sensation for various ground solar reflectance in climate chamber. The asterisk ** means t-test ($P < 0.05$).

ground surfaces, and a significant difference was shown between the white sheet and the other two ground surfaces. The mean values of the reported thermal sensation on the concrete and black sheet were different, but the difference was not significant. This confirmed that the greater the solar reflectance, the greater the heat that the human body received on the ground surface with different amounts of solar reflectance; the human thermal load also increased, consequently affecting the bodily thermal sensation. There is a substantial difference based on the amount of reflected solar radiation. The lower the amount of solar reflectance, the greater the surface temperature increase and the larger the amount of infrared radiation from the ground surface. The influence of reflected solar radiation was substantial, and the greater the amount of solar reflectance, the greater the net radiation, thereby increasing the human thermal load.

The human subject experiments were conducted simultaneously in an open space with direct sun exposure and at a location under a neighboring tree canopy that blocked the solar radiation. The dates were August 2 and 3, 10 and 11, 13 and 14, 20 and 21, and 27 and 28, at five locations with a variable leaf area index (LAI). Six subjects took part between 11:00 and 14:00. With two subjects used as a pair, three pairs were tested per day. The subjects switched the locations on the second day. The weather on the experiment days was calm and clear. The air temperature and humidity were almost the same on the 2 days of the experiments.

The tree canopy reduced both the amount of solar radiation from above, and the amount of infrared radiation, which was influenced by the reduction in ground surface temperature. The mean values of the reported thermal sensation are also different, as shown in Fig. 5.12, and the difference is significant. Fig. 5.13 shows the correlation between the human thermal load and thermal sensation. The correlation coefficient was $R^2 = 0.354$. The vertical and horizontal axes show the differential of the human thermal load and thermal sensation based on the subject pair (tested

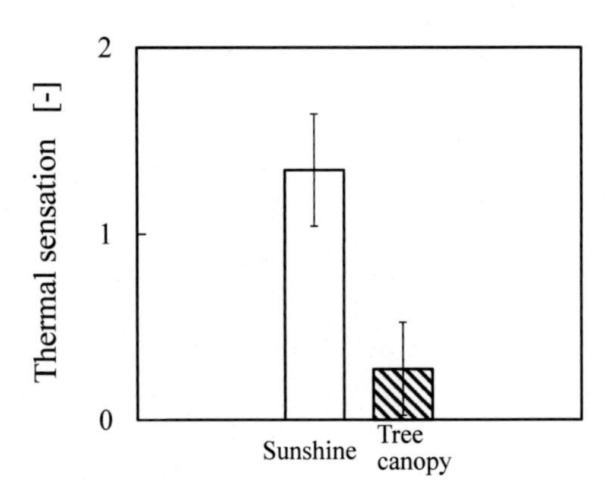

FIGURE 5.12

Thermal sensation under tree canopy.

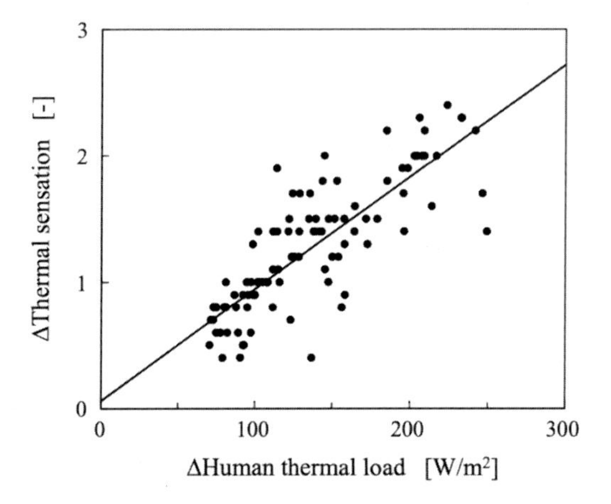

FIGURE 5.13

Correlation in differentials between human thermal load and human thermal sensation under tree canopy.

simultaneously) for each reported time. When the test subjects were exposed to solar radiation, the human thermal load increased, as did the reported thermal sensation values. The human thermal load is lower under the tree canopy at the corresponding reported time.

5. An evaluation method based on numerical analysis

5.1 A method for evaluating thermal environment in outdoor spaces

In this section, we outline an evaluation method based on the numerical analysis. Fig. 5.14 outlines the computational method used for evaluating the thermal environment in outdoor spaces. This method is composed of the following three steps: (1) setting initial and boundary conditions, (2) calculation of spatial distributions of thermal environmental elements by using the computational fluid dynamics (CFD) analysis coupled with convection and radiation, and (3) evaluation of thermal comfort for pedestrians in outdoor spaces. The initial and boundary conditions for the calculation at the step (2) are determined using the input data. These include meteorological and geometrical data and surface conditions. The CFD analysis using boundary conditions determined from the input data is carried out to examine the spatial distributions of thermal environmental elements, such as wind velocity, air temperature, radiation, and humidity. The thermal comfort in the outdoor spaces is evaluated from the results of calculations by using human thermoregulation models based on the results of the CFD analysis. The present method enables us to obtain the distributions of the SET*, WBGT, and universal thermal comfort index (UTCI) [17] as the thermal comfort scale.

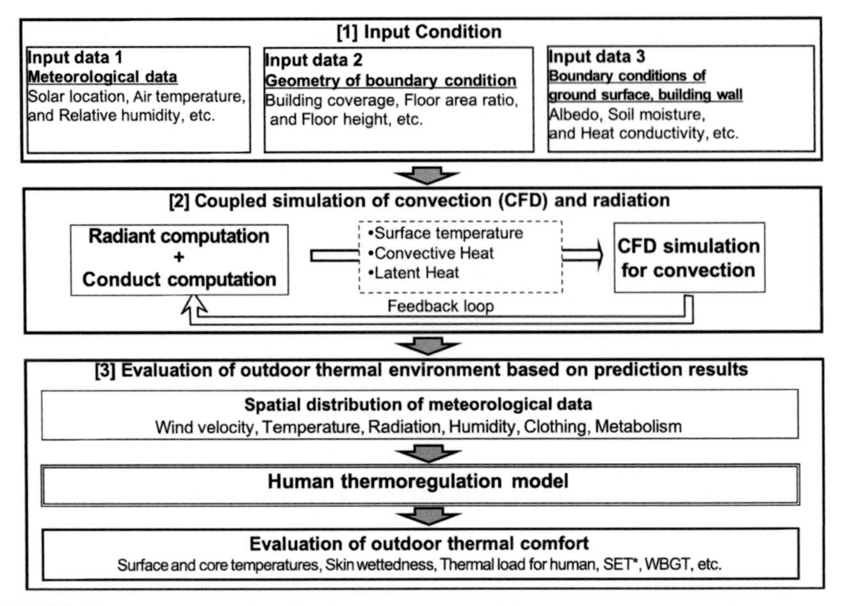

FIGURE 5.14

Flowchart for assessing outdoor thermal comfort based on computational fluid dynamics (CFD).

5.2 A method for radiant analysis in outdoor spaces

For the evaluation of the thermal environment in the outdoor spaces, it is required to examine the distributions of the heat exchanges of both the solar and the long-wave radiations between the surface elements comprising the computational domain. Fig. 5.15 illustrates an example of the coordinate system in an outdoor space for the radiant computation. The surfaces of the obstacles and the space are divided into small grids in order to consider the radiant exchanges between complex terrain

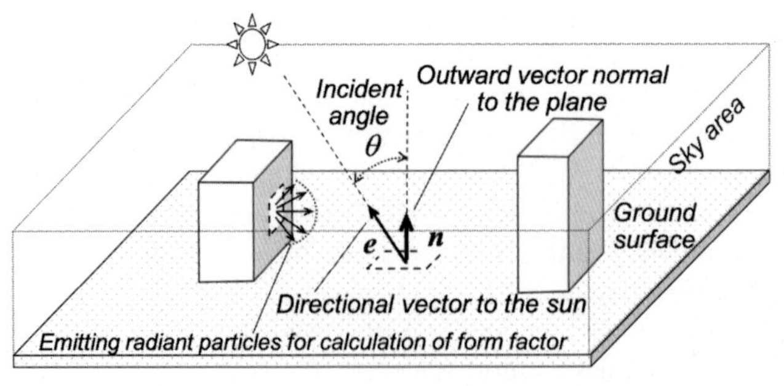

FIGURE 5.15

Outdoor coordinate system for the radiant computation.

and geometries. The physical properties and temperature on each surface element are assumed to be uniform in this computation.

As mentioned at the next section, the distributions of the surface temperature at each surface element are obtained using a heat budget equation at the surface of the ground, outside a building, and on the tree crown. For example, equations for the absorbed solar radiation gain of surface element i comprising the computational domain S_i (W) are expressed as follows:

$$S_i = (\alpha_{i\theta} E_{Di} + \alpha_i E_{Si}) + \alpha_i \left(\sum_{j=1}^{N} F_{ji} R_{Sj} \right), \tag{5.10}$$

$$R_{Si} = (\rho_{i\theta} E_{Di} + \rho_i E_{Si}) + \rho_i \left(\sum_{j=1}^{N} F_{ji} R_{Sj} \right), \tag{5.11}$$

$$E_{Di} = A_i \gamma_i I_N \cos \theta_i, \tag{5.12}$$

$$E_{Si} = A_i F_{iS} I_{SH}, \tag{5.13}$$

where A_i is the area of i and E_{Di} and E_{Si} are the direct and sky solar radiation gains to grid i (W), respectively. Furthermore, F_{ij} is the form factor, i.e., the fraction of radiation leaving i that is intercepted by j; F_{iS} is the form factor from i to the sky; I_N is the direct solar radiation incident on the normal surface (W/m^2); I_{SH} is the sky solar radiation incident on the horizontal surface (W/m^2); R_{Si} is the short-wave radiosity or the total short-wave radiation energy flux of a surface per unit time at i (W); S_{Ti} is the transmitted solar radiation at i (W); γ_i is the irradiation ratio at i; and θ_i is the incident angle of the sun's rays to the plane. The symbols α_i, ρ_i, and τ_i denote the absorptance, reflectance, and transmittance of solar radiation on i, respectively, while $\alpha_{i\theta}$, $\rho_{i\theta}$, and $\tau_{i\theta}$ those at the incident angle of the sun's rays to the plane (θ). The symbols α_i, ρ_i, τ_i, $\alpha_{i\theta}$, $\rho_{i\theta}$, and $\tau_{i\theta}$ also include the ratio of window area to the surface area. In these equations, the multiple reflections are considered by corrections of the R_{Si} updated with iterative calculations based on the progressive radiosity method [18,19]. For the computations of these radiation gains, it is necessary to also obtain the distributions of F_{ij}, F_{iS}, and γ_i. In the present method, these values are obtained by using the computational method based on the Monte Carlo method [20].

5.3 Calculation of heat balance on ground surface and outside a building

The distributions of the surface temperature T_i at each point are obtained using a heat balance equation at the surface of the ground and outside a building. Let S_i, R_i, H_i, D_i, and LE_i be the absorbed solar radiation gain of surface i (W), long-wave radiation gain of i (W), convection heat transmission at i (W), conduction to the building or ground (W), and heat dissipation by evaporation from i (W), respectively. The heat balance equation consists of these elements, as shown in Eq. (5.14):

$$S_i + R_i + H_i + D_i + LE_i = 0. \tag{5.14}$$

On the left side of Eq. (5.14), a positive value indicates an inflow of energy to i, while a negative value indicates an outflow of energy. The equations for calculating each item in Eq. (5.14) are expressed as follows.

R_i (W) is calculated using the following equations:

$$R_i = -E_i + \varepsilon_i \sum_{j=1}^{N} F_{ji} R_{Lj}, \tag{5.15}$$

$$R_{Li} = E_i + (1 - \varepsilon_i) \sum_{j=1}^{N} F_{ji} R_{Lj}, \tag{5.16}$$

$$E_i = A_i \bullet \varepsilon_i \bullet \sigma T_i^4, \tag{5.17}$$

where E_i is the long-wave radiation emitted at i (W), R_{Li} is a long-wave radiosity at i (W), ε_i is the absorption rate of long-wave radiation of i, and σ is the Stephan–Boltzmann constant (W/[m^{2}K^4]) (= 5.67×10^{-8} W/[m^{2}K^4]).

H_i is calculated using Eq. (5.18).

$$H_i = A_i \cdot \alpha_C (T_{ai} - T_i), \tag{5.18}$$

where T_{ai} is the air temperature in the region adjacent to i (K), and α_C is the convective heat transfer coefficient (W/[m^{2}K]).

D_i is calculated using Eq. (5.19).

$$D_i = -A_i \cdot \lambda (T_i - T_{bi}) / \Delta z, \tag{5.19}$$

where λ is the heat conductivity of the building material or ground (W/[m K]), and T_{bi} is the inside wall or underground temperature at depth Δz, derived by solving the transient heat conduction problem in the solid.

LE_i is calculated using Eq. (5.20).

$$LE_i = A_i \cdot L \cdot \alpha_W \beta_i (pa_i - pSat_i), \tag{5.20}$$

where pa_i is the water vapor pressure at the region adjacent to i (kPa), $pSat_i$ is the saturated water vapor pressure at i (kPa), L is the latent heat of evaporation (J/kg), α_W is the moisture transfer coefficient (kg/[m^2 s kPa]), and β_i is the moisture availability at i.

5.4 Convection calculation based on CFD analysis

For the examination of the spatial distributions of the thermal environmental elements in outdoor spaces, it is required that a suitable turbulence model is applied in the CFD analysis. In the present method, the standard k−ε model [21] is used as the fundamental turbulence model. Table 5.2 summarizes the governing equations of the standard k−ε model. The standard k−ε model is a well-known turbulence model for conveniently analyzing the airflow in spaces and is utilized to examine the thermal environment and the air pollution in the indoor space. However, it has been also pointed out that this model has the following issues concerning the accuracy of the computational results: (1) the lack of the consideration of the effects of buoyancy on the turbulent-flow heat flux and (2) the overestimation of the turbulent energy at the windward side of a bluff body [22]. Hence, in the present method, we incorporated the following two effects into the standard k−ε model: (1) the effect of buoyancy on the evaluation of the turbulent-flow heat flux as summarized in Table 5.3 [23,24] and (2) inclusion of the function of the modified Kato−Launder model [25] that regulates excessive production of the turbulent-flow energy k on the building windward side as summarized in Table 5.4. The revised k−ε model was utilized in the analyses described in the following Sections 5.6 and 5.7 in this chapter.

5.5 Numerical modeling of effects of planting trees on thermal environment in outdoor spaces

Planting trees is most familiar measure incorporated into the planning or the design of the adaptive cities against the hot environment. Hence, it is absolutely necessary to incorporate a numerical modeling of effects of the trees on the thermal environment in the outdoor spaces. The computational method described in the present chapter includes the numerical plant canopy model developed by Yoshida et al. [26]. This numerical model considers the following three effects of planted trees on the thermal environment in the outdoor spaces: (1) drag of planted trees, (2) shading the solar and the long-wave radiations, and (3) transpiration of water vapor from the plant canopy. Fig. 5.16 illustrates the conceptual diagram of the plant canopy model described here.

The strongest advantage of the plant canopy model introduced here is to enable us to three-dimensionally consider effects of planted trees on the thermal environment in the urban area. The conventional models including these effects of planted trees were mainly developed in the field of agricultural meteorology [27,28]. In these models, horizontal distributions of physical elements were assumed to be uniform. However, it is not adequate to use these plant canopy models for evaluating the urban microscale climate because spatial distributions of wind, temperature, radiation, and humidity in the urban space are highly three-dimensional. Hence, it is obvious that the plant canopy model introduced here is more suitable to evaluate effects of planted trees on the thermal environment in the outdoor spaces than the conventional models. In this section, concepts of this plant canopy model are outlined below.

Table 5.2 Governing equations of the standard k-ε model.

1) Equation of continuity

$$\frac{\partial\langle u_i\rangle}{\partial x_i}=0 \quad (5.21)$$

2) Equation of motion

$$\frac{\partial\langle u_i\rangle}{\partial t}+\langle u_j\rangle\frac{\partial\langle u_i\rangle}{\partial x_j}=-\frac{\partial\langle p\rangle}{\partial x_i}+\frac{\partial}{\partial x_j}\left[\nu_t\left(\frac{\partial\langle u_i\rangle}{\partial x_j}+\frac{\partial\langle u_j\rangle}{\partial x_i}\right)\right]-g_i\beta(\langle\theta\rangle-\theta_0) \quad (5.22)$$

3) Equation of turbulent energy k

$$\frac{\partial k}{\partial t}+\langle u_i\rangle\frac{\partial k}{\partial x_i}=\frac{\partial}{\partial x_i}\left[\frac{\nu_t}{\sigma_k}\frac{\partial k}{\partial x_i}\right]+P_k+G_k-\varepsilon \quad (5.23)$$

4) Equation of turbulent dissipation rate ε

$$\frac{\partial\varepsilon}{\partial t}+\langle u_i\rangle\frac{\partial\varepsilon}{\partial x_i}=\frac{\partial}{\partial x_i}\left[\frac{\nu_t}{\sigma_\varepsilon}\frac{\partial\varepsilon}{\partial x_i}\right]+\frac{\varepsilon}{k}(C_{\varepsilon1}P_k+C_{\varepsilon3}G_k)-C_{\varepsilon2}\frac{\varepsilon^2}{k} \quad (5.24)$$

5) Transport equation of air temperature

$$\frac{\partial\langle\theta\rangle}{\partial t}+\langle u_i\rangle\frac{\partial\langle\theta\rangle}{\partial x_i}=\frac{\partial}{\partial x_i}\{-\langle u_i'\theta'\rangle\} \quad (5.25)$$

6) Transport equation of absolute humidity

$$\frac{\partial\langle q\rangle}{\partial t}+\langle u_i\rangle\frac{\partial\langle q\rangle}{\partial x_i}=\frac{\partial}{\partial x_i}\{-\langle u_i'q'\rangle\} \quad (5.26)$$

7) Subequations

$$P_k=\nu_t S^2 \quad (5.27)$$

$$G_k=-g_3\beta\langle u_3'\theta'\rangle \quad (5.28)$$

$$\nu_t=C_\mu\frac{k^2}{\varepsilon} \quad (5.29)$$

$$S=\sqrt{\frac{1}{2}\left(\frac{\partial\langle u_i\rangle}{\partial x_j}+\frac{\partial\langle u_j\rangle}{\partial x_i}\right)^2} \quad (5.30)$$

$$\langle u_i'\theta'\rangle=-\frac{\nu_t}{\sigma_\theta}\frac{\partial\langle\theta\rangle}{\partial x_i} \quad (5.31)$$

$$\langle u_i'q'\rangle=-\frac{\nu_t}{\sigma_W}\frac{\partial\langle q\rangle}{\partial x_i} \quad (5.32)$$

$g_3=-9.8$, $C_{\varepsilon1}=1.44$, $C_{\varepsilon2}=1.92$, $C_{\varepsilon3}=1.44$ ($G_k>0$), 0 ($G_k\leq 0$), $\sigma_k=1.0$, $\sigma_\varepsilon=1.3$, $\sigma_\theta=0.9$, $\sigma_W=0.9$

Table 5.3 Equations of turbulent heat flux Including buoyant effect.

1) Turbulent heat flux

$$\langle u_3'\theta'\rangle = -\frac{\nu_t}{\sigma_\theta}\frac{\partial\langle\theta\rangle}{\partial x_3} - \frac{k}{\varepsilon}C_{\theta 3}g_3\beta\langle\theta'^2\rangle \quad (5.33)$$

2) Transport equation of temperature variance $<\theta'^2>$

$$\frac{\partial\langle\theta'^2\rangle}{\partial t} + \langle u_i\rangle\frac{\partial\langle\theta'^2\rangle}{\partial x_i} = -2\langle u_i'\theta'\rangle\frac{\partial\langle\theta\rangle}{\partial x_i} + \frac{\partial}{\partial x_i}\left[\frac{\nu_t}{\sigma_{\theta'}}\frac{\partial\langle\theta'^2\rangle}{\partial x_i}\right] - 2\frac{\varepsilon}{k}\frac{\langle\theta'^2\rangle}{2R} \quad (5.34)$$

$C_{\theta 3} = 0.25$, $R = 0.8$, $\sigma_\theta = 0.5$, $\sigma_{\theta'} = 1.0$

5.5.1 Drag force of planted trees

The numerical plant canopy model incorporated into the computational method introduced here considers the following two aerodynamic effects of planted trees: (1) the effects of planted trees on decrease of wind velocity and (2) the effects of planted trees on the amount of increase in both turbulence and energy dissipation rate. Here, the occupancy of fluid is assumed to be 1.0, even in the mesh where tree exists, because the volume of trees is negligibly small compared with the mesh volume.

Table 5.4 Equations of modified Kato–Launder model.

1) Kato–Launder model

$$P_k = \nu_t S\Omega \quad (5.35)$$

$$\nu_t = C_\mu\frac{k^2}{\varepsilon} \quad (5.36)$$

$$\Omega = \sqrt{\frac{1}{2}\left(\frac{\partial\langle u_i\rangle}{\partial x_j} - \frac{\partial\langle u_j\rangle}{\partial x_i}\right)^2} \quad (5.37)$$

$$S = \sqrt{\frac{1}{2}\left(\frac{\partial\langle u_i\rangle}{\partial x_j} + \frac{\partial\langle u_j\rangle}{\partial x_i}\right)^2} \quad (5.38)$$

2) Modified Kato–Launder model

$$P_k = \nu_t S\,\Omega (\text{in case of } \Omega/S \le 1) \quad (5.39)$$

$$P_k = \nu_t S^2 (\text{in case of } \Omega/S > 1) \quad (5.40)$$

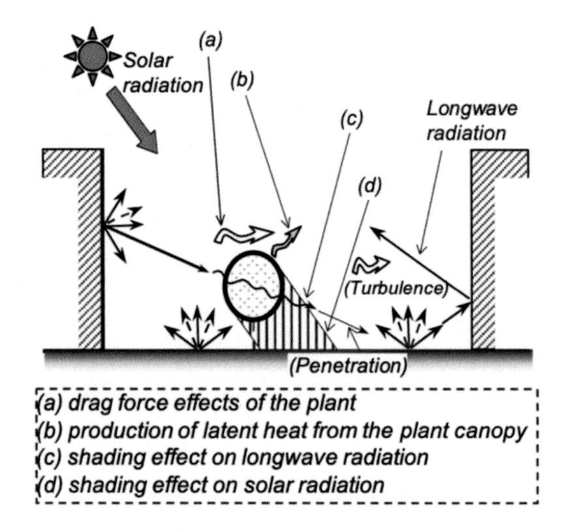

FIGURE 5.16

Conceptual diagram on effects of planting a tree on the thermal environment in an outdoor space.

The effects of planted trees on decrease of velocity are given by adding the negative value of the drag force of planted trees "$-F_i$" as the extra term in the momentum equation (Eq. 5.22). F_i is calculated by the following equation:

$$F_i = \eta \cdot C_f \cdot a_d \cdot \langle u_i \rangle \sqrt{\langle u_i \rangle^2}, \tag{5.41}$$

where η is the fraction of the area covered with trees, C_f is the drag coefficient for tree canopies, and a_d is the leaf area density (1/m).

The effects of planted trees on the amount of increase in the turbulence are simulated by including the extra term F_k in the transport equation of the turbulent energy k (Eq. 5.23). The equation of the F_k is expressed as follows:

$$F_k = \langle u_i \rangle \cdot F_i, \tag{5.42}$$

The effects of planted trees on the amount of increase in the energy dissipation rate are also simulated by including the extra term F_ε in the transport equation of the turbulent dissipation rate ε (Eq. 5.24). The equation of the F_ε is expressed as follows:

$$F_\varepsilon = \frac{\varepsilon}{k} \cdot C_{p\varepsilon 1} \cdot F_k, \tag{5.43}$$

where $C_{p\varepsilon}$ is the model coefficient and is set to be 1.8 in the present analyses described in this chapter [29].

5.5.2 Shading effects of the solar and the long-wave radiations by the plant canopy

As mentioned in Section 5.2 in this chapter, the radiant heat transport is calculated by using the method based on the Monte Carlo method. In this calculation, it is assumed that both the solar and the long-wave radiant fluxes incident to the plant canopy are decayed at the rate expressed by the following model function f_{DP}:

$$f_{DP} = 1 - \exp(k' \cdot a_d \cdot l_p), \tag{5.43}$$

where k' is the extinction coefficient for the solar and the long-wave radiations, and l_P is the length of the trajectory passed through the plant canopy by the radiant flux.

5.5.3 Transpiration of water vapor from the plant canopy

In the plant canopy model described here, the amount of transpiration of the water vapor from the plant canopy E_P is obtained by solving the heat budget equation on the leaf surface of the plant canopy. The equations for the heat budget on the leaf surface of the plant canopy P is expressed as follows:

$$S_P + R_P + H_P + LE_P = 0, \tag{5.44}$$

$$H_P = A_P \cdot \alpha_C (T_{aP} - T_P), \tag{5.45}$$

$$LE_P = A_P \cdot \alpha_W \cdot \beta_P \cdot L(pa_P - pSat_P), \tag{5.46}$$

where S_P is the absorbed solar radiation gain of the leaf surfaces of the plant canopy P (W), R_P is the long-wave radiation gain of P (W), H_P is the convection heat transmission at P (W), and LE_P is the heat dissipation by evaporation from P (W), respectively. Furthermore, T_P is the surface temperature of P (K); T_{aP} is the temperature of the air included in P (K); pa_P is the water vapor pressure of the air included in P (kPa); $pSat_P$ is the saturated water vapor pressure of the air included on P (kPa); and β_P is the moisture availability at P. Symbol A_P denotes the amount of the leaf area in the tree crown (m^2) and equals twice the product of the volume of tree crown V_P and the leaf area density of tree crown a_d. In these equations, the heat conduct term is neglected because the heat capacity of each leaf in the tree crown is negligibly small. On the left side of Eq. (5.44), a positive value indicates an inflow of energy to P, while a negative value an outflow of energy.

The effects of planted trees on decrease in the air temperature are considered by including the extra term Q_{HP} in the transport equation of the air temperature θ (Eq. 5.25). The equation of the Q_{HP} is expressed as follows:

$$Q_{HP} = -\frac{H_P}{\rho C_P \cdot V_P}, \tag{5.47}$$

where ρ is the air density (kg/m^3), C_P is the specific heat of the air (J/[kg K]), and V_P is the volume of the cell including the tree crown (m^3).

The effects of planted trees on increase in the absolute humidity are also considered by including the extra term Q_{EP} in the transport equation of the absolute humidity q (Eq. 5.26). The equation of the Q_{EP} is expressed as follows:

$$Q_{EP} = -\frac{LE_P}{\rho L \cdot V_P}, \tag{5.48}$$

5.6 Validation of the evaluation method [30]

5.6.1 Outline of the analysis

In this section, we introduce a study of a numerical analysis for the validation of the evaluation method described at the previous sections. Fig. 5.17 illustrates both the plan and the photo of the study area for the present analysis. The study area was an outdoor space within a courtyard canyon space in Akabane-dai housing development in North ward in Tokyo. In the study area, the following two computational domains were established: Domain A and Domain B. Domain A was established at the center of the north–west area of the study area, while Domain B covered almost whole of the study area. Domain A was established to validate the accuracy of the computational results from the evaluation method. Yoshida et al. [31] measured the thermal environment here in the summer season during the period from August 2 to 6 in 1999 and obtained detailed observation data. Hence, we compared the computational results with the observation data in order to validate the evaluation method. On the other hand, Domain B was established for improving the accuracy of the computational results in Domain A. The computational results in Domain A are strongly affected by the boundary conditions of the thermal environmental elements such as wind velocity vectors, air temperature, and humidity. Then, in the present analysis, we performed the numerical analyses with the following procedure. At first, we carried out the numerical analysis of the flow field in the Domain B by using the observation data of the meteorological conditions of the target period and obtained the spatial distributions of both the wind velocity vectors and the turbulence. After the first computation, a part of the computational results in Domain B was imposed on the boundary conditions for the computation in Domain A. The boundary data in the domain A except of the flow field were set to be the measurement data homogeneously. The thermal properties for the building walls and the ground, such as thermal conductivity, albedo, and emissivity, were determined by reference to results of investigations of the structures of both the building walls

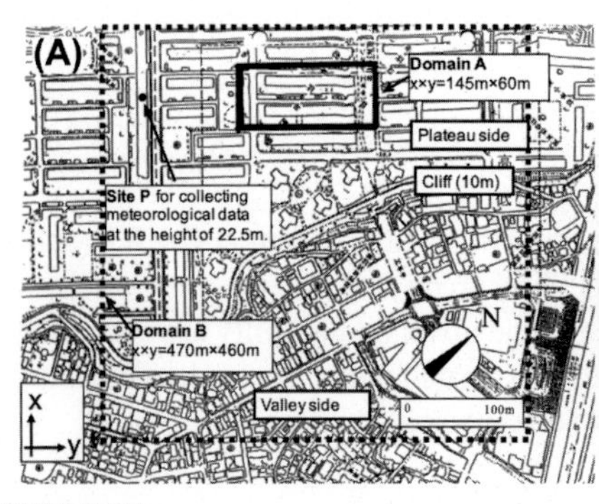

FIGURE 5.17

Study area for the numerical analysis for the validation of the evaluation method. (A) Plan; (B) photo of the courtyard canyon space in Domain A.

and the ground materials. The properties concerning the trees in Domain A, such as the leaf area density and the drag force coefficient, were set to be suitable values based on species of the actual planted trees. The meteorological conditions were observed on the roof top of a building (Site P) that was approximately 22.5m tall, as shown in Fig. 5.17.

The target period was set to be a clear sunny day in summer. The analysis starts from 6:00 on August 3, and a time integration of 42 h was performed using the meteorological data. The validation of the evaluation method was carried out by comparing the computational results on August 4 with the measurement data.

5.6.2 Comparison of ground surface temperature between the computational results and the measurement data

Fig. 5.18 illustrates the diurnal variations of the ground surface temperature at the following area: the sunny area covered with the grass surface (Point 1), the shade area covered with the grass (Point 2), and the sunny area paved with asphalt (Point 3). The computational results on each point corresponded well with the measurement data. In particular, both the computational results and the measurement data at Point 1 ranged from approximately 26°C to approximately 42°C, so the difference was significantly small.

Fig. 5.19 also illustrates the comparison of distributions of the grand surface temperature between the computational results and the measurement data. The

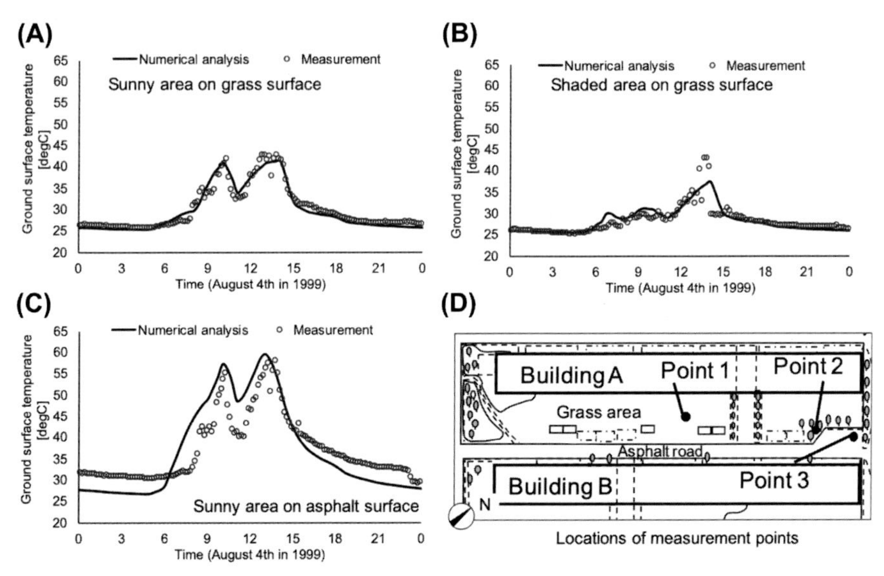

FIGURE 5.18

Diurnal variations of the ground surface temperature on August 4th in 1999. (A) Variations at the sunny area covered with the grass surface (Point 1); (B) Variations at the shade area covered with the grass area (Point 2); (C) Variations at the sunny area paved with asphalt (Point 3); (D) Locations of measurement points.

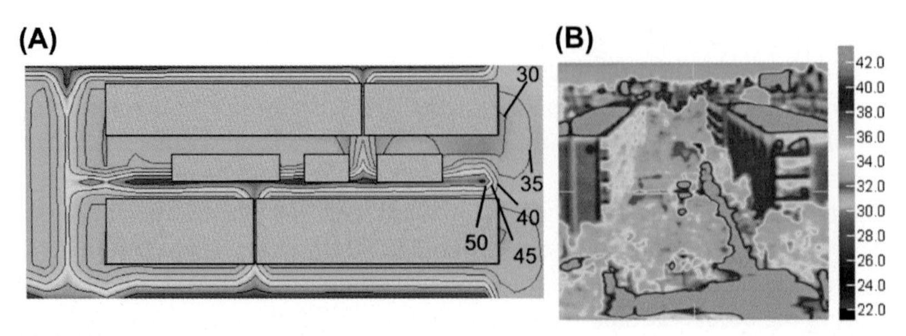

FIGURE 5.19

Comparison of distributions of the grand surface temperature at 15:00 August 4th in 1999 between the computational results and the measurement data. (A) Distributions of the grand surface temperature from the computational results. (B) A picture taken with an infrared radiation thermometer from Site P for collecting the meteorological data at the height of 22.5 m.

measurement data were obtained by taking a picture image with an infrared radiation thermometer from Site P where the meteorological data were observed. In the computational results, the value in the sunny area paved with asphalt was approximately 50°C, while the values in both the grass and the shaded area were approximately 30°C. A large difference between these sites was seen from the computational results. These tendencies were also seen in the measurement data. Hence, the computational results agreed well with the measurement.

5.6.3 Comparison of flow field between the computational results and the measurement data

Fig. 5.20 illustrates the comparison of the flow field between the computational results and the measurement data. In this comparison, the horizontal distributions of

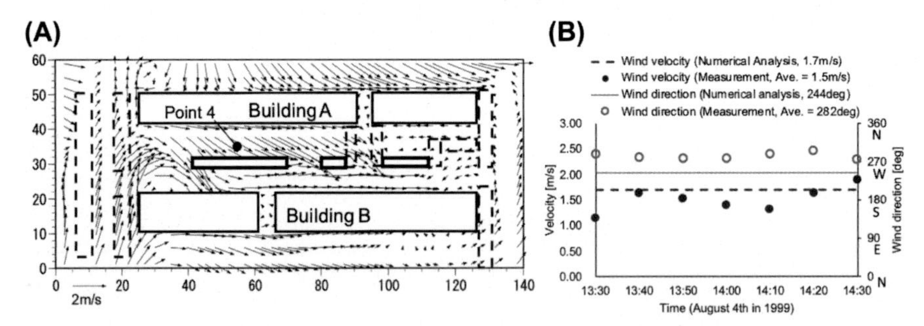

FIGURE 5.20

Comparison of the flow field between the computational results and the measurement data. (A) Horizontal distributions of the wind velocity vectors at the height of 1.5 m at 14:00 August 4th in 1999, by the computational results. (B) Temporal variations of both the wind direction and the wind velocity in the period from 13:30 to 14:30 on Point 4 by the measurement data. Point 4 was located on the center part in the courtyard canyon space.

the wind velocity vectors at the height of 1.5 m at 14:00 on August 4 were referred as the computational results. On the other hand, we also referred the temporal variations of both the wind direction and the wind velocity in the period from 13:30 to 14:30 on Point 4 located on the center part in the courtyard canyon space as the measurement data. The measurement data were obtained by using a three-dimensional ultrasonic anemometer. In the computational results, the wind direction and the wind velocity on Point 4 were approximately SW and approximately 1.7 m/s, respectively. On the other hand, the measurement data averaged in the period were approximately W and approximately 1.5 m/s, respectively. From the comparison, it was thought that the computational results correspond well with the measurement data although the differences of the wind direction were slightly seen.

5.6.4 Comparison of air temperature between the computational results and the measurement data

Fig. 5.21 illustrates comparison of the horizontal distributions of air temperature at the height of 1.5 m at 14:00 on August 4 between the computational results and the measurement data. The values from the computational results ranged from approximately 32°C to 37°C, while the measurement data from approximately 33 to 35°C. Hence, the differences between the computational results and the measurement data were quite small. The values on the area shaded by trees ranged from approximately 32 to 33°C and were approximately 1°C lower than those on the surroundings. This was caused by the fact that the sensible heat releases from the ground surface near the trees were reduced by effects of the tree crowns shading the solar radiation.

5.6.5 Conclusion of the study for the validation

In this section, we introduced an example for examining the validation of the computational method for evaluating the thermal environment in the adaptive cities. From the investigation, it was clarified that the validity of the proposed method was confirmed because the results of the proposed method agreed well with the measurement data.

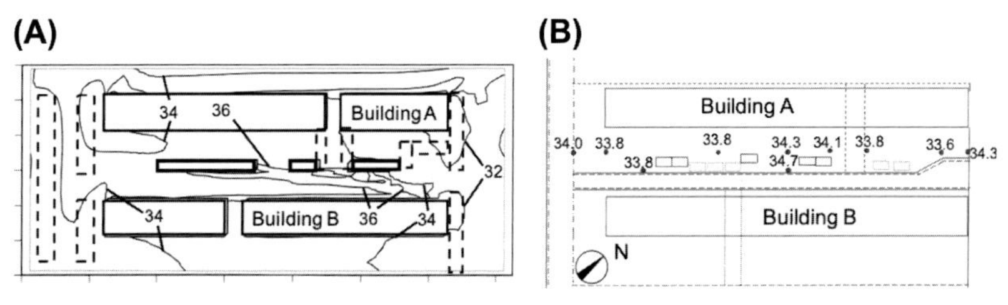

FIGURE 5.21

Comparison of the horizontal distributions of air temperature at the height of 1.5 m at 14:00 August 4th in 1999 between the computational results and the measurement data. (A) Computational results; (B) measurement data.

5.7 Applications of the evaluation method [26]

5.7.1 Introduction

In the previous section, we outlined the study for the validation of the computational method for evaluating the thermal environment in the adaptive cities. In this section, we introduce a study on examinations of effects of both scales and types of greening on the thermal comfort for pedestrians in the summer season as an application example of the proposed method for the evaluation of installing the countermeasure techniques to the urban planning.

As described in Section 5.5 in this chapter, greening is most familiar measure incorporated into the planning or the design of the adaptive cities against the hot environment. Building designers and urban planners have often utilized greening such as planting trees, covering the ground with grass, and making the garden, in order to improve the thermal environment in the outdoor spaces. However, both scales and types of greening significantly affect the thermal environment in the outdoor space. Additionally, mechanisms of effects of greening on the thermal environment are significantly complicated. Fig. 5.22 illustrates conceptual diagram for effects of trees on the thermal comfort for pedestrians in the summer season. The tree has various influences as follows: (1) shading the solar radiation, (2) reducing the reflected heat, (3) emitting the latent heat, (4) decrease of the air temperature, (5) decrease of the wind velocity, and (6) increase of the humidity. In these influences, the items (1), (2), (3), and (4) contribute to improve the thermal comfort for the pedestrians in the summer season, while the items (5) and (6) worsen it.

Effects of trees	Changes of environmental elements	Thermal comfort for pedestrians in summer
Shading solar radiation	Decrease of MRT	Improved
Decrease of longwave radiation	Decrease of MRT	Improved
Production of latent heat	Decrease of air temperature	Improved
Drag force effect	Decrease of wind velocity	Degraded
Transpiration of water vapor	Increase of relative humidity	Degraded

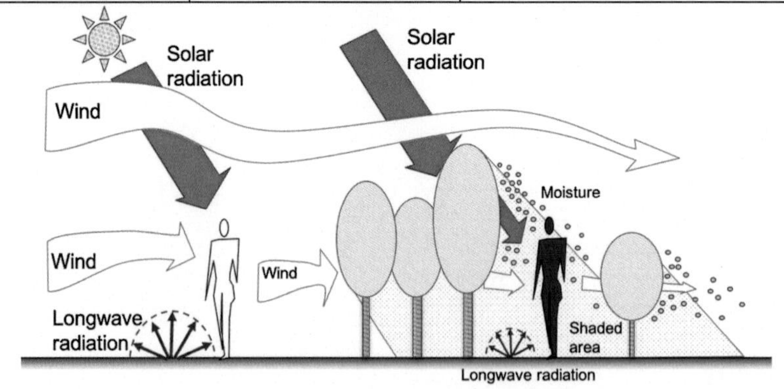

FIGURE 5.22

Conceptual diagram for effects of planted trees on the thermal comfort for pedestrians in the summer season.

Hence, it has been understood that planting trees does not always improve the thermal comfort for pedestrians in the summer season because the negative effects are included. Hence, it is necessary to evaluate quantitatively both the positive and the negative effects of trees on the thermal comfort for the pedestrian. In this section, we introduce a study on the quantitative evaluation on effects of various types of green on the thermal environment in the outdoor spaces using the proposed method.

5.7.2 Outline of analysis

Fig. 5.23 illustrates the computational domain in the present analysis. The 30-m cubic-shaped building was set on regularly to the computational domain. The geometries of these blocks were modeled by referring to data for the Ginza area of Tokyo. One unit of the block area is selected for the analysis in order to reduce the computational load by decrease of the total mesh number. The periodic boundary conditions are imposed to the side boundaries of the computational domain. The target date was decided to be July 23 in order to investigate the thermal environment during a typical hot summer day. Meteorological data measured by the Japan Meteorological Agency in Tokyo were used in this study. The thermal comfort for the pedestrians was evaluated using the results obtained at 15:00 on the target date. The meteorological conditions of air temperature, wind velocity, wind direction, relative humidity, sun's azimuth, sun's altitude, amount of the direct solar radiation to the normal plane, amount of the sky solar radiation to the horizontal plane, and amount of downward atmospheric radiation at 15:00 are assumed to be, respectively, 31.6°C, 3 m/s, S, 58%, 83.2 degrees, 45.2 degrees, 731 W/m^2, 136 W/m^2, and 424 W/m^2. Here, both the wind velocity and the wind direction were observed at the height of 74.6 m.

Three computational cases were considered in the present analysis. Table 5.5 summarizes it, and the conceptual figures for it were shown in Fig. 5.24. In case 1, it is assumed that the grass area ratio, or the ratio of area covered with grass to the whole ground surface area, was set to be 10%, and the ground surface except

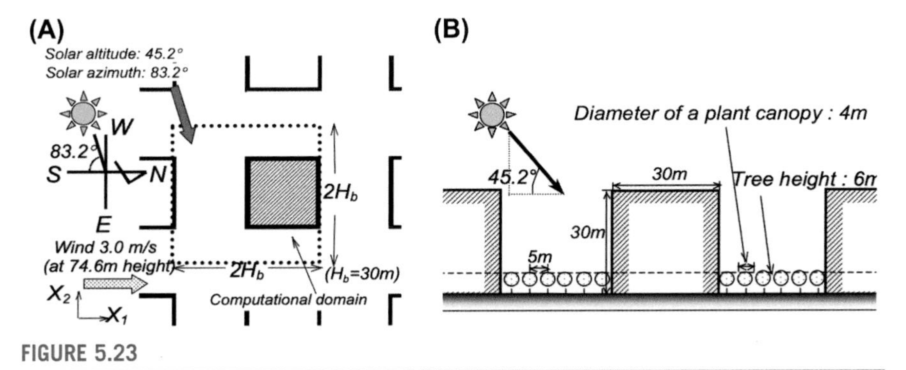

FIGURE 5.23

Computational domain for the quantitative evaluation on effects of various types of green on the thermal environment in the outdoor spaces. (A) Plan; (B) section in case 3.

Table 5.5 Computational cases.

	Grass area ratio of ground surface (%)	Plant canopy
case1	10	Without plant canopy
case2	100	
case3		With plant canopy

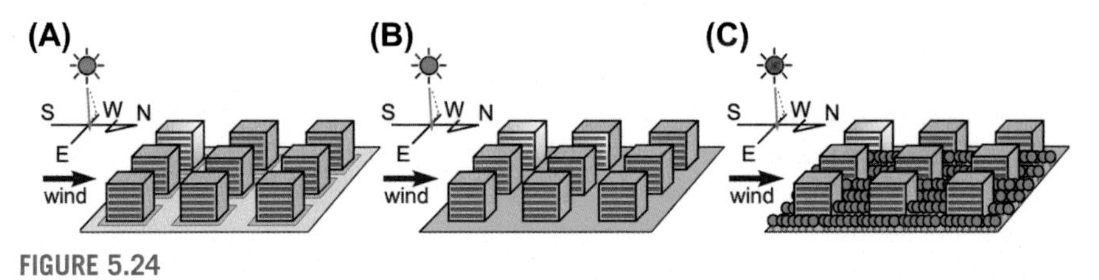

FIGURE 5.24

Conceptual diagram for computational cases. (A) Case 1 (grass area ratio: 10%, without plant canopies); (B) case 2 (grass area ratio: 100%, without plant canopies); (C) case 3 (grass area ratio: 100%, with plant canopies).

for the grass area was paved with asphalt. In case 2, it was assumed that the whole ground surface was covered with grass. Furthermore, in case 3, trees with crown of 4m diameters are supposed to be planted on the same ground surface of case 2 at regular intervals of 5 m. Table 5.6 summarizes the properties of planted trees.

5.7.3 Distributions of ground surface temperature

Fig. 5.25 illustrates the distributions of the ground surface temperature. The sun's azimuth and altitude were 83.2 degrees (nearly west) and 45.2 degrees. The building height also equals to the distance between each building. In cases 1 and 2, the sunny area was seen in a west—east street while the shaded area was seen in the street canyon wedged between the east-west side building walls. Thus, the significant

Table 5.6 Properties of planted trees.

Properties	Values
Tree species	Gingko
Tree height (m)	6.00
Diameter of a tree crown (m)	4.00
Height of dead branch base (m)	2.00
Leaf area density a_d (m^2/m^3)	1.00
Drag force coefficient C_d (-)	0.20
Extinction coefficient k' (-)	0.60
Moisture availability of plant canopy β_P (-)	0.30
Albedo (-)	0.15
Emissivity (-)	0.98

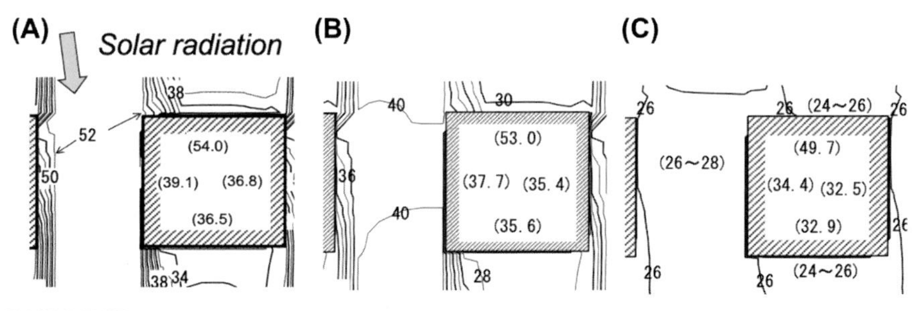

FIGURE 5.25

Distributions of the ground surface temperature. (A) Case 1 (grass area ratio: 10%, without plant canopies); (B) case 2 (grass area ratio: 100%, without plant canopies); (C) case 3 (grass area ratio: 100%, with plant canopies).

differences of the values between the sunny area and the shaded one were observed in cases 1 and 2. However, the values in case 3 ranged from approximately 24°C to approximately 28°C, and the difference of the temperature between the sunny area and the shaded one becomes smaller compared to the other cases. This was caused by the fact that the values in the sunny area decreased by shading effects of plant canopies in case 3. Hence, it was found that planting trees significantly affects the distributions of both the sunny and the shaded area. The values of the ground surface temperature in the sunny area were approximately 52°C in case 1, 40°C in case 2, and 28°C in case 3. Thus, it was found that effects of increasing grass area were approximately 12°C, while those of planting trees were approximately 14°C in the sunny area.

5.7.4 Distributions of wind velocity vectors

Fig. 5.26 illustrates the horizontal distributions of the wind velocity vectors at the height of 1.5 m. In cases 1 and 2, it was seen that the wind blew along the street in a north–south direction. We also saw two horizontal circulations in the street

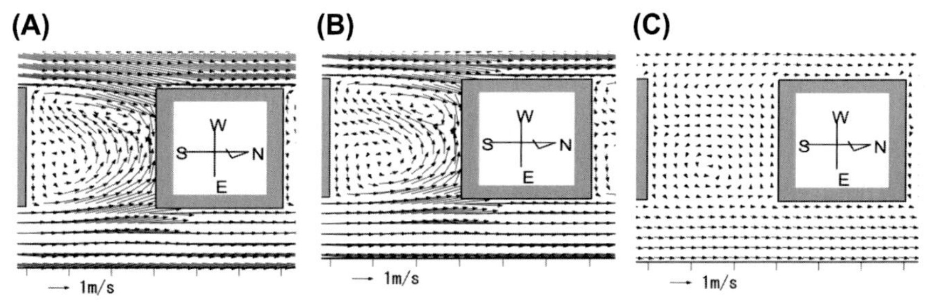

FIGURE 5.26

Horizontal distributions of the wind velocity vectors at the height of 1.5m. (A) Case 1 (grass area ratio: 10%, without plant canopies); (B) case 2 (grass area ratio: 100%, without plant canopies); (C) case 3 (grass area ratio: 100%, with plant canopies).

canyon. The difference of the results between cases 1 and 2 is not so large. On the other hand, the values of the wind velocity in case 3 are fairly lower than those in case 2. This reduction of the wind velocity is attributed to the effects of the drag force of planted trees on the whole area in the outdoor spaces.

5.7.5 Distributions of air temperature

Fig. 5.27 illustrates the horizontal distributions of the air temperature at the height of 1.5 m. The values in case 1 ranged from approximately 31.5 to 34°C, while in case 2 from approximately 31.0 to 32.5°C. Furthermore, the values in case 3 also ranged from approximately 27.0 to 29.0°C. Hence, it was found that increasing the grass area reduced the air temperature by from approximately 0.5 to 1.5°C, while planting trees by from approximately 2.0 to 5.0°C.

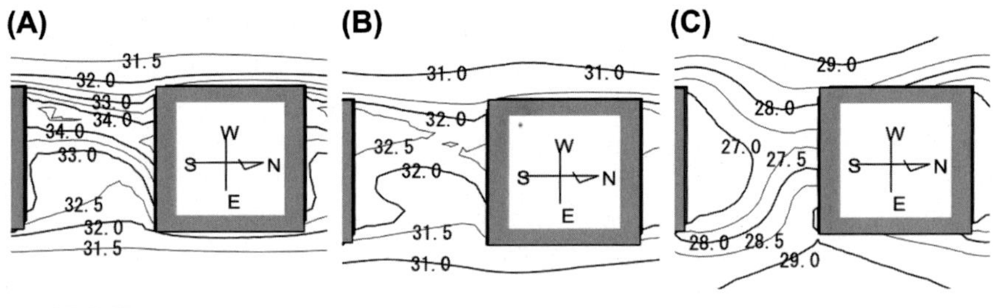

FIGURE 5.27

Horizontal distributions of the air temperature at the height of 1.5 m. (A) Case 1 (grass area ratio: 10%, without plant canopies); (B) case 2 (grass area ratio: 100%, without plant canopies); (C) case 3 (grass area ratio: 100%, with plant canopies).

5.7.6 Distributions of relative humidity

Fig. 5.28 illustrates the horizontal distributions of the relative humidity at the height of 1.5 m. The values in case 1 ranged from approximately 51% to approximately

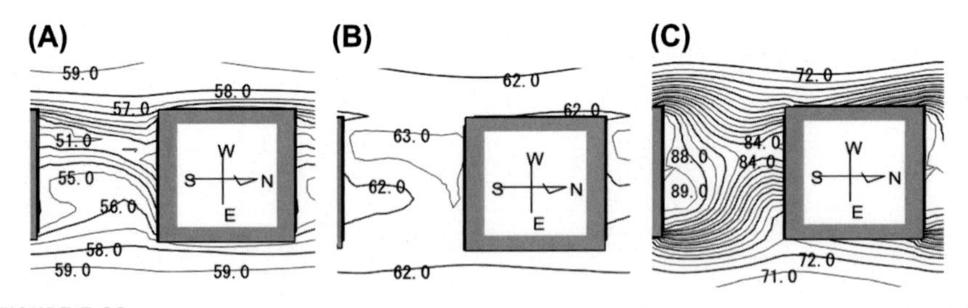

FIGURE 5.28

Horizontal distributions of the relative humidity at the height of 1.5 m. (A) Case 1 (grass area ratio: 10%, without plant canopies); (B) case 2 (grass area ratio: 100%, without plant canopies); (C) case 3 (grass area ratio: 100%, with plant canopies).

59%, while in case 2 from approximately 62% to approximately 63%. Furthermore, the values in case 3 also ranged from approximately 71% to approximately 89%. The values in case 2 were approximately from 3% to 8% higher than those in case 1, due to effects of both the increase of the absolute humidity by generating the water vapor from the grass surface and the decrease in the air temperature. Furthermore, the values in case 3 were approximately from 10% to 20% higher than those in case 2. This difference was caused by effects of both the increase of the absolute humidity with generating the water vapor from the trees and the decrease of the air temperature in case 3.

5.7.7 Distributions of mean radiant temperature

Fig. 5.29 illustrates the horizontal distributions of the mean radiant temperature (MRT) for the entire body of a pedestrian. In the present analysis, the MRT was calculated using a prismatic (specifically, a microcube) human body model as follows:

$$\sigma T_{mrt}^4 = \sum_{l=1}^{6} [(\alpha_l S_l + R_l) \cdot c_l], \tag{5.49}$$

where T_{mrt} is the MRT for the entire body of a pedestrian (K); the subscript l is the index to express the direction of the surface, which is composed of a microcube (six surfaces in total); S_l is the incident solar radiation on surface l of the microcube (W/m^2); R_l is the incident longwave radiation to l (W/m^2); and c_l is the weighting coefficient related to the absorption of solar radiation by each surface of the microcube. The coefficient 0.024 was used for the top and bottom surfaces of the cube, and 0.238 was used for the sides [32]. The symbol a_l is the absorptance of the short-wave radiation for the human body. In the present analysis, the values of the absorptance of the short-wave radiation were set as 0.5. The value 0.5 is used under the assumption that the pedestrian is wearing clothing of the style common for a man in an office building during the summer season.

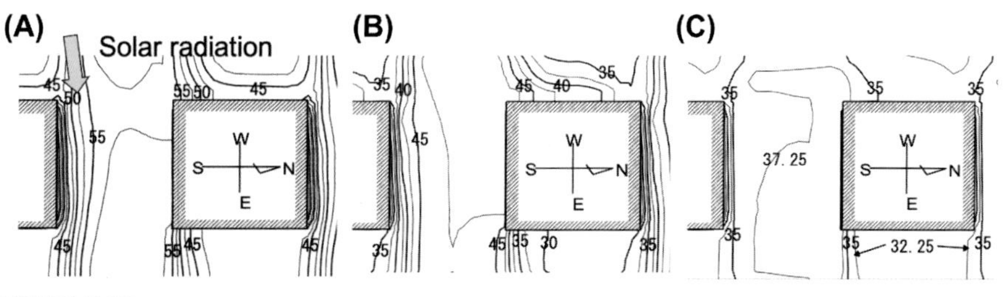

FIGURE 5.29

Horizontal distributions of mean radiant temperature (MRT) for the entire body for the pedestrians. (A) Case 1 (grass area ratio: 10%, without plant canopies); (B) case 2 (grass area ratio: 100%, without plant canopies); (C) case 3 (grass area ratio: 100%, with plant canopies).

The values in both the sunny and the shade area in case 1 were approximately 55 and 45°C, while in case 2 approximately 45 and 30°C, respectively. The values in both the sunny and the shade area in case 2 were approximately 10°C higher than those in case 1 due to the decrease of the ground surface temperature with increasing the grass area. On the other hand, the values in case 3 ranged from approximately 33 to 38°C, and the difference of MRT between the sunny and the shaded areas becomes smaller compared to the other cases. This is caused by the same reason for the results of the ground surface temperature described above. Hence, it was found that both increasing the grass area and planting trees reduced MRT by approximately 10°C.

5.7.8 Distributions of SET*

Fig. 5.30 illustrates the horizontal distributions of the new standard effective temperature (SET*) for the entire body of a pedestrian. For the calculation of the physiological responses described in Section 3.2 in this chapter, such as the mean skin temperature (θsk), the mean skin wettedness (w_{sk}), and the total heat flux at the skin surface (Q_{sk}), the clothing insulation and the metabolic heat generation were assumed to be 0.5 clo and 1.2 met, respectively. In case 1, the values in sunny areas ranged from approximately 32 to 44°C, and those in shaded areas from approximately 30 to 32°C. Thus, we saw significant differences between both areas. This tendency was also seen in case 2. However, this difference was very small in case 3 due to the effects of shading the solar radiation by planted trees. Fig. 5.31 (1) shows the change of SET* by the increase in the grass area ratio (SET* in case 2 minus SET* in case 1), and the change of SET* by planting trees (case 3 minus case 2) is given in Fig. 5.31 (2). In these figures, gray zone corresponds to the negative value indicating the area where SET* decreases by increasing the green area and thus the thermal environment in the summer season is improved. Fig. 5.31 (1) shows that the values of SET* decreased on almost the whole domain except for the small area in the recirculating region by 0.1–2.0°C with increasing

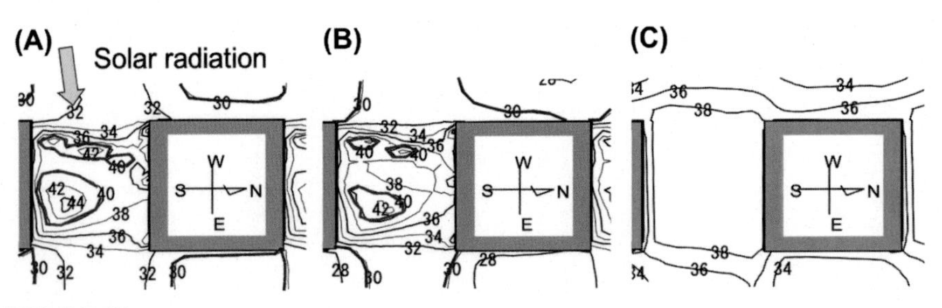

FIGURE 5.30

Horizontal distributions of standard effective temperature (SET*) for the entire body for the pedestrians. (A) Case 1 (grass area ratio: 10%, without plant canopies); (B) case 2 (grass area ratio: 100%, without plant canopies); (C) case 3 (grass area ratio: 100%, with plant canopies).

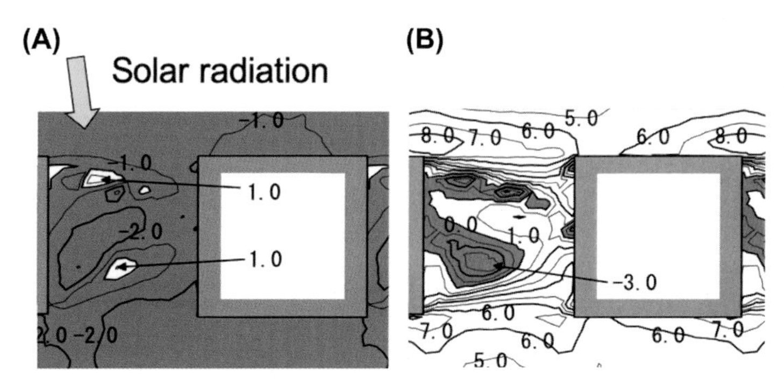

FIGURE 5.31

Horizontal distributions of changes of standard effective temperature (SET*) by greening. (A) Change of SET* by increasing the grass area (the values in case 2 minus those in case 1); (B) change of SET* by planting trees (the values in case 3 minus those in case 2). Gray zone indicates the area where the values of SET* decrease by increasing the green area and thus the thermal environment in the summer season is improved.

grass area (case 1 → case 2). On the other hand, white zone, where SET* increase, is wider than the gray zone in Fig. 5.31 (2). This increase of the white zone means that planting trees caused a rising discomfort in the thermal environment in the outdoor spaces in the summer season. These results reflect the fact that the wind velocity of case 3 was greatly decreased, as a density of trees in planting area was too much. It indicates that the effect of wind velocity on the outdoor thermal environment is significantly large. Thus, the suitable arrangement of planted trees is required to improve the pedestrian comfort in outdoor space.

5.7.9 Conclusion of the application study

The effects of various types of green were investigated using the evaluation method proposed here. In the results of this study, it was found that increasing grass area ratio improves thermal environment in outdoor space during the summer season, while planting trees does not necessarily improve the thermal environment because of reduction of wind velocity due to effects of drag force by plant canopies. Future direction of this study is to clarify the suitable density of planting trees.

6. Summary

In this chapter, we outlined methods to evaluate effects of installing the countermeasure techniques to these practical planning on the pedestrians in outdoor spaces.

At the former part of this chapter, we outlined the relationship between the elements concerning the thermal environment and the responses form the pedestrians. From the descriptions, it was understood that various physical and mental elements as well as the air temperature affect the thermal comfort for the pedestrians. After that, the basic theory of the thermoregulation based on the heat budget of the human body was introduced because the thermal comfort for the pedestrians has a strong relationship to the thermoregulation mechanisms in the human body. In this part, we also outlined the human thermoregulation model as tools for the estimation of the various physiological mechanisms in the human body.

At the latter part of this chapter, we introduced the method for evaluating the thermal environment in the adaptive cities. The evaluation scales of the adaptive cities were outlined at the beginning of this part. In this part, the WBGT and the SET* were introduced as the primary evaluation scales for the evaluations of the thermal environment in the outdoor space. After that, the evaluation methods based on the field observation were outlined. In the present method, the thermal environment in the outdoor spaces was evaluated by using the human thermal load that was derived from the heat budget in the human body. Finally, we also introduced the evaluation method based on the numerical analysis. The present method was based on CFD analysis coupled with both the radiant computation and the human thermoregulation computation. The validation study and the application example were also outlined in the last part of this chapter.

From the descriptions above, the outlines of the evaluation methods of the adaptive cities were clarified. It was understood that the evaluation and the examination based on the heat budget in the human body are required at the investigations by using both the field measurement and the numerical analysis because the heat budget significantly affects the thermal comfort for the pedestrians. The future agenda on these evaluation methods is further investigations by using the field measurement and the numerical analysis for improving the validity of the evaluation.

List of symbols

$<f>$	time averaged value of f
a_d	leaf area density (1/m)
A_i	area of surface element i (m^2)
A_P	amount of the leaf area in the tree crown P (m^2)
B	amounts of the heat loss via the blood flow from the core to the skin layers (W/m^2)
C	amounts of the convective heat exchange between the skin layer and the surroundings (W/m^2)
C_f	drag coefficient for tree canopy
c_l	weighting coefficient related to the absorption of solar radiation by each surface of the microcube
C_P	specific heat of the air (J/[kg K])
C_{Pl}	drag coefficient for tree canopy
$C_{P\varepsilon l}$	model coefficient (= 1.8)
D	amounts of the heat loss via the heat conduct from the core to the skin layers (W/m^2)
D_i	heat conduction to the building or ground at surface element i (W)

E_{Di}	direct solar radiation gains to surface element i (W)
E_i	long-wave radiation emitted at surface element i (W)
E_{Si}	sky solar radiation gains to surface element i (W)
ET^*	new effective temperature (°C)
f_{cl}	clothing area factor
F_{cl}	intrinsic clothing thermal efficiency
f_{cl}	clothing area factor
f_{DP}	damping rate of both solar and the long-wave radiant fluxes incident to the plant canopy
F_i	drag force of planted trees (m/s^2)
F_{ij}	form factor, i.e., the fraction of radiation leaving surface element i that is intercepted by j
F_{iS}	form factor from surface element i to the sky
F_k	the extra term in the transport equation of the turbulent energy k (m^2/s^3)
F_{load}	human thermal load (W/m^2)
F_{pcl}	permeation efficiency
f'	deviation from $<f>$, $f' = f - <f>$
F_ε	extra term in the transport equation of the turbulent dissipation rate ε (m^2/s^4)
H_i	convection heat transmission at surface element i (W)
H_P	convection heat transmission at plant canopy P (W)
I_N	direct solar radiation incident on the normal surface (W/m^2)
I_{SH}	sky solar radiation incident on the horizontal surface (W/m^2)
k	turbulent energy (Table 5.2) (m^2/s^2)
k'	extinction coefficient for the solar and the long-wave radiations
l	index to express the direction of the surface
L	latent heat of evaporation (J/kg)
LE	Amounts of the evaporative heat exchange between the skin layer and the surroundings (W/m^2)
LE_i	Heat dissipation by evaporation from surface element i (W)
LE_P	Heat dissipation by evaporation from plant canopy P (W)
l_P	Length of the trajectory passed through the plant canopy by the radiant flux (m)
M	Amount of metabolic heat production (W/m^2)
OT	Operative temperature (°C)
p	Pressure (Table 5.2) (Pa)
pa	Vapor pressure in ambient air (kPa)
pa_i	Water vapor pressure at the region adjacent to surface element i (kPa)
P_k	Production term of k (m^2/s^3)
$pSat_i$	Saturated water vapor pressure at surface element i (kPa)
$pSat_P$	Saturated water vapor pressure of the air included on plant canopy P (kPa)
$pSat_{sk}$	Saturated water vapor pressure at the skin surface (kPa)
q	Absolute humidity (Table 5.2) (kg/kg [DA])
Q_{EP}	Extra term in the transport equation of the absolute humidity q (kg/ (kg(DA) s)
Q_{HP}	Extra term in the transport equation of the air temperature θ (K/s)
Q_{RES}	Total heat flux at the skin surface (W/m^2)
Q_{sk}	Amount of the heat exchange from respiration (W/m^2)

r	Solar reflectance
R	Amounts of the radiant heat exchange between the skin layer and the surroundings (W/m^2)
R_i	Long-wave radiation gain of surface element i (W)
R_l	Incident long-wave radiation to surface l of the microcube (W/m^2)
R_{Li}	Long-wave radiosity or the total long-wave radiation energy flux of a surface per unit time at surface element i (W)
R_{net}	Net radiation (W/m^2)
R_P	Long-wave radiation gain of the leaf surfaces of the plant canopy P (W)
R_{Si}	Short-wave radiosity or the total short-wave radiation energy flux of a surface per unit time at surface element i (W)
S	Heat storages per the skin area in the human body (W/m^2)
S_{cr}	Heat storages per the skin area in the core layer (W/m^2)
$SET*$	New standard effective temperature (°C)
S_i	Absorbed solar radiation gain of surface element i (W)
S_l	Incident solar radiation on surface l of the microcube (W/m^2)
S_P	Absorbed solar radiation gain of the leaf surfaces of the plant canopy P (W)
S_{sk}	Heat storages per the skin area in the skin layer (W/m^2)
S_{Ti}	Transmitted solar radiation at surface element i (W)
T_{ai}	Air temperature in the region adjacent to surface element i (K)
T_{aP}	Temperature of the air included in plant canopy P (K)
T_{bi}	Inside wall or underground temperature at depth Δz of surface element i (K)
T_D	Dry-bulb temperature (°C)
T_G	Globe temperature (°C)
T_i	Surface temperature at surface element i (K)
T_{mrt}	Mean radiant temperature (MRT) for the entire body of a pedestrian (K)
T_P	Surface temperature of plant canopy P (K)
T_W	Naturally ventilated wet-bulb temperature (°C)
u_i	Three components of the velocity vector
V_P	Volume of tree crown P (m^3)
W	Amount of mechanical work accomplished (W/m^2)
$WBGT$	Wet-bulb globe temperature (°C)
w_{sk}	Mean skin wettedness
x_i	Three components of spatial coordinate ($i = 1, 2, 3$: streamwise, lateral, vertical) (m)
β	Coefficient of volumetric expansion (1/K)
ε	Dissipation rate of k (Table 5.2) (m^2/s^3)
θ	Temperature (Table 5.2) (°C)
λ	Heat conductivity of the building material or ground (W/[m K])
ρ	Air density (Table 5.2) (kg/m^3)
η	Fraction of the area covered with trees
σ	Stephan–Boltzmann constant ($W/[m^2K^4]$) ($= 5.67 \times 10^{-8}$ $W/[m^2K^4]$)
α_C	Convective heat transfer coefficient ($W/[m^2K]$)
α_e	Latent heat transfer coefficient at the mean skin area ($W/[m^2kPa]$)
α_i	Absorptance of solar radiation on surface element i
β_i	Moisture availability surface element i
ρ_i	Reflectance of solar radiation on surface element i
γ_i	Irradiation ratio at surface element i

ε_i	Emissivity of long-wave radiation at surface element i
θ_i	Incident angle of the sun's rays to the plane i
τ_i	Transmittance of solar radiation on surface element i
$\alpha_{i\theta}$	Absorptance of solar radiation on surface element i at the incident angle of the sun's rays to the plane θ
$\rho_{i\theta}$	Reflectance of solar radiation on surface element i at the incident angle of the sun's rays to the plane θ
$\tau_{i\theta}$	Transmittance of solar radiation on surface element i at the incident angle of the sun's rays to the plane θ
β_P	Moisture availability at plant canopy P
θ_{sk}	Mean skin temperature (°C)
α_t	Sensible heat transfer coefficient at the mean skin area (W/[m^{2}K])
ν_t	Eddy viscosity (Table 5.2) (m^2/s)
α_W	Moisture transfer coefficient (kg/[m$^2\cdot$s$\cdot$kPa])

References

[1] Fanger PO. Thermal comfort. New York: McGraw-Hill; 1970.

[2] Gagge AP, Stolwijk JAJ, Nishi Y. A standard predictive index of human response to the thermal environment. ASHARE Transactions 1986;92(1):709–31.

[3] Stolwijk JAJ. A mathematical model of physiological temperature regulation in man. NASA contractor report-1885. 1971.

[4] Gordon RG, Roemer RB, Horvath SM. A mathematical model of the human temperature regulatory system e transient cold exposure response. IEEE Transactions on Biomedical Engineering 1976;23:434–44.

[5] Smith CE. A transient, three-dimensional model of the human thermal system. Kansas State University; 1993. Dissertation.

[6] Fu G. A transient, 3-D mathematical thermal model for the clothed human. Kansas State University; 1995. Dissertation.

[7] Huizenga C, Zhang H, Arens E. A model of human physiology and comfort for assessing complex thermal environments. Building and Environment 2001;36:691–9.

[8] Tanabe S, Kobayashi K, Nakano J, Ozeki Y, Konishi M. Evaluation of thermal comfort using combined multi-node thermoregulation (65MN) and radiation models and computational fluid dynamics (CFD). Energy and Buildings 2002;34:637–46.

[9] Tanabe S, Kobayashi K, Ogawa K. Development of numerical thermoregulation model COM for evaluation of thermal environment. Journal of Architecture, Planning and Environmental Engineering (Transactions of AIJ) 2006;599:31–8 [in Japanese with English abstract].

[10] Sato T, Xu L, Tanabe S. Development of human thermoregulation model JOS applicable to different types of human body. In: Proceedings of healthy buildings 2003, Singapore, 7–11 December 2003, 1, 828–834; 2003.

[11] Kobayashi M, Tanabe S. Development of JOS-2 human thermoregulation model with detailed vascular system. Building and Environment 2013;66:1–10.

[12] Kinoshita S, Yoshida A, Shimazaki Y. Measurement of heat transfer and evaluation of thermal comfort in urban street area – measurement in summer season at Midosuji Osaka city. AIJ Journal of Technology and Design 2009;15(No.31):803–6.

[13] JIS Z8504:1999 (ISO 7243: 1989), Hot environments - estimation of the heat stress on working man, based on the WBGT-index (Wet blub globe temperature).

[14] Hardy JD, Dubois EF. The technic of measuring of radiation and convection. The Journal of Nutrition 1938;15:461–75.

[15] Shimazaki Y, Yoshida A, Kinoshita S. Proposal on thermal comfort index based on human thermal load. Transactions of JSRAE 2009;26(1):113–20.

[16] Yoshida A, Hayashi D, Yasuhiro, Shimazaki Y, Kinoshita S. Evaluation of thermal sensation in various outdoor radiation environments. Architectural Science Review 2019. https://doi.org/10.1080/00038628.2019.1597678.

[17] Jendritzky G, Dear RD, Havenith G. UTCI—why another thermal index? International Journal of Biometeorology 2012;56:421–8.

[18] Chen SE. Incremental radiosity: an extension of progressive radiosity to an interactive image synthesis system. Computer Graphics 1990;24(4):135–44.

[19] Shao MZ, Badler NI. A gathering and shooting progressive refinement radiosity method. Technical Reports (CIS). USA: Department of Computer and Information Science, University of Pennsylvania; 1993.

[20] Oomori T, Taniguchi H, Kudo K. Monte Carlo simulation of indoor radiant environment. International Journal for Numerical Methods in Engineering 1990; 30(4):615–27.

[21] Launder BE, Spalding DB. The numerical computation of turbulent flows. Computer Methods in Applied Mechanics and Engineering 1974;3:269–89.

[22] Murakami S, Mochida A, Hayashi Y. Examining the k-ε model by means of a wind tunnel test and large-eddy simulation of the turbulence structure around a cube. Journal of Wind Engineering and Industrial Aerodynamics 1990;35:87–100.

[23] Launder BE. On the computation of convective heat transfer in complex turbulent flows. Transactions of the ASME, Journal of Heat Transfer 1988;110:1112–28.

[24] Noguchi Y, Murakami S, Mochida A, Tominaga Y. Numerical study on thermal environment in urban area. Part 3. Incorporation of buoyancy effect into modelling for $\langle u'_3\theta' \rangle$ in k-ε model. In: Annual technical meeting of architecture institute of Japan, Japan, D; 1994. p. 65–6 [in Japanese].

[25] Kato M, Launder BE. The modelling of turbulent flow around stationary and vibrating square cylinders. In: Prep. of the 9th symposium on turbulent shear flow, Kyoto, Japan, 16–18 August; 1993. p. 1–6.

[26] Yoshida S, Ooka R, Mochida A, Murakami S, Tominaga Y. Development of three dimensional plant canopy model for numerical simulation of outdoor thermal environment. In: Sixth international conference on urban climate (ICUC6), Göteborg, Sweden; 2006. p. 320–3.

[27] Wilson NR, Shaw RH. A higher order closure model for canopy flow. Journal of Applied Meteorology 1977;16:1197–205.

[28] Yamada T. A numerical model study of turbulence airflow in and above a forest canopy. Journal of the Meteorological Society of Japan 1982;60:439–54.

[29] Mochida A, Tabata Y, Iwata T, Yoshino H. Examining tree canopy model for CFD prediction of wind environment at pedestrian level. Journal of Wind Engineering and Industrial Aerodynamics 2008;96:1667–77.

[30] Harayama K, Yoshida S, Ooka R, Mochida A, Murakami S. Prediction of outdoor environment with unsteady coupled simulation of convection, radiation and conduction Part 1, Numerical study based on unsteady radiation and conduction analysis. Journal of

Architecture, Planning and Environmental Engineering (Transactions of AIJ) 2002;556: 99−106 [in Japanese with English abstract].

[31] Yoshida S, Murakami S, Narita K, Takahashi T, Ooka R, Mochida A, Tominaga Y. Field measurement of outdoor thermal environment within courtyard canyon space around apartment complex in summer. Journal of Architecture, Planning and Environmental Engineering (Transactions of AIJ) 2002;552:69−76 [in Japanese with English abstract].

[32] Nakamura Y. Expression method of the radiant field on a human body in buildings and urban spaces. Journal of Architecture, Planning and Environmental Engineering (Transactions of AIJ) 1987;561:21−9 [in Japanese with English abstract].

CHAPTER

The role of local government

6

Keiko Masumoto, Dr.

Researcher, Research Group, Osaka City Research Center of Environmental Science, Osaka, Japan

Chapter outline

1. Introduction

Over the years, climate change has worsened worldwide, and because of its large spatiotemporal scale, it is difficult for people to recognize climate change in their daily lives. It was feared decades ago that global warming would go slowly, so that even if it reached a serious situation, it would not be able to return, even if it did not notice a small change in temperature.

In recent years, however, damage caused by torrential rains and typhoons has increased, and it has become ever more urgent to not only prevent further global warming but also take prompt measures to minimize the damage.

The change in the thermal environment is not just a global problem; it cannot be ignored from a regional perspective either. Apart from the global warming phenomenon, the "urban heat island phenomenon," which is creation of a "heat island" as the temperature of the city rises significantly compared to its surrounding areas, also cannot be ignored. Higher temperatures in such areas are believed to increase the negative impact on people's health, besides global warming. These may be due to topography and atmospheric pressure characteristics, such as the Foehn phenomenon, but largely, higher temperatures in cities are a result of man-made factors,

Adaptation Measures for Urban Heat Islands. https://doi.org/10.1016/B978-0-12-817624-5.00006-3

such as land cover with few green spaces, urban structures with dense buildings, and increased energy consumption. The national and local governments are working on measures to be taken to control these factors.

2. Administrative procedures

In order to formulate measures and promote effective countermeasures, the government performs the following procedures.

(1) Understanding the actual conditions of a hot environment: In the case of local governments, it is thought that a hot environment is also characterized by factors that are different from those in other cases, such as topography, land use, and economic activity. Therefore, first and foremost, it is necessary to grasp the actual situation.

(2) Understanding health risks: It is necessary to explain in advance to the citizens how and to what extent it will affect them. Regarding the cost of the measures, citizens need to form a consensus; health risks, in particular, are a great concern.

(3) The "Measure plan for heat island" created by the government shows the menu and schedule of measures, then promises publicly that specific measures will actually be implemented. The government needs to proceed with "Plan-Do-Check-Act (PDCA)" which creates a plan, implements measures, verifies the effect, and takes action toward a new step for the next plan.

(4) Setting goals: There is a need to set numerical goals, not just effort goals. There is also a need to provide some support and a rationale for achieving the goals.

(5) Policy menu selection and effect estimation: Each measure has a menu with multiple items, which are shown so that people can choose the right method in the right place. In addition, the effect achieved by introducing each measure is shown.

(6) Implementation of countermeasures—as an administration, operator, and citizen: Some private companies introduce advanced measures in line with government plans, even if they are not shown in these plans. In addition, the government aims to increase the number of examples of countermeasures and spread them far and wide by considering budgets and subsidies for "model projects" as advanced methods.

(7) Understanding the progress of measures: While measures are being taken in each direction according to the plan, it is indispensable to grasp the progress status in order to achieve the goal achievement level.

(8) Verification of effects: After a certain target period has passed, we will verify whether the countermeasures have been effective and whether the plan target has been achieved, which will serve as the basis for considering its next stage.

(9) Dissemination and awareness to promote the plan: Public administration not only makes plans but also implements "model projects" and expects businesses and citizens to voluntarily work on countermeasures. These procedures promote measures to improve a hot environment and reduce health risks with the agreement of business operators and citizens.

3. Health risks due to temperature rise

An increase in the number of transported patients due to heat stroke has been reported by local governments—because the 2018 summer in Japan was extremely hot, the number of emergency cases from May to September was 95,137 nationwide (The preliminary figures, Ministry of Internal Affairs and Communications), and 2259 people in Osaka City with a population of 2.7 million (according to Table 6.1: Osaka City Fire Bureau data), the number of heatstroke patients per million population exceeded 800, more than twice that of 2017. This number does not include cases where patients received medical care directly, took appropriate measures and escaped, or died. It is presumed that there are more occurrences of heat stroke in places where people gather outdoors, such as event venues, amusement parks, and sports stadiums, besides the poor living on the streets and residents living without air conditioners.

In particular, there are many elderly people living alone in the city, and even if the room temperature rises, they cannot feel themselves, so appropriate room temperature management cannot be performed. It is no exaggeration to say that patients with heart disease or infants who do not have sufficient body temperature regulation can be at risk of life even if they live in normal living spaces in a hot city.

Table 6.1 Increment of heat stroke patients transported to hospitals in Osaka City.

	Number of patients			Patients per million
Year	Male	Female	Total	
2002	137	47	184	70.4
2003	105	21	126	48.1
2004	135	66	201	76.6
2005	129	43	172	65.4
2006	173	67	240	91.1
2007	230	109	339	128.3
2008	308	105	413	155.8
2009	144	49	193	72.6
2010	684	300	984	369.2
2011	502	255	757	283.5
2012	510	238	748	279.4
2013	799	472	1271	473.6
2014	497	259	756	281.4
2015	697	449	1146	425.8
2016	736	416	1152	425.7
2017	690	381	1071	394.7
2018	1281	978	2259	832.4

The resident population in Osaka City is about 2.7 million people.

4. Detailed air temperature monitoring

Under such circumstances, when taking measures, it is necessary to first understand the extent and location of temperature changes occurring. A large city has a regional meteorological observatory, and the temperature as well as various observations, which are representative, are made public. These are typical observation values in a wide area including the city, and if they are used as indicators, they can be used when considering comprehensive measures in a wide area. However, typical observations from weather stations are not sufficient to take concrete measures in places where the temperature is high and the risk of heat stroke is high, such as in densely populated areas or where people gather. There is a need to compare whether the temperature in a narrow area with high risk is higher than other areas, and then, investigate the change with time, i.e., grasp the "spatiotemporal characteristics" of the temperature and reflect it in specific measures.

In Osaka City, which has an area if 225 km^2, there is only one regional meteorological observatory. However, measurements by local government were taken at 70 locations and the temperature distribution in the city area was observed in detail—during the day, the temperature was higher on the inland side compared to the temperature in the harbor where sea breeze is likely to enter, but the city center was relatively hot at night (Figs. 6.1 and 6.2). These results indicate that there is a need to develop distinctive measures for each region.

5. Dissemination of information on the heat stroke risk using the heat index (wet-bulb globe temperature)

The Ministry of Environment (Japan) uses the wet-bulb globe temperature (WBGT) calculated from dry-bulb temperature, wet-bulb temperature, and black-bulb temperature as an index of heat stroke risk. To make the risk known, we measured it in various places in major cities and made predictions in advance. The risk from this indicator has been reported in conjunction with weather forecasts on television and has been used in schools to limit outdoor activities and events. This indicator that has been publicized is a value that should be used as a representative value in the region. However, as it should originally be measured at each location, portable measuring instruments have been mass-produced, and observation at schools and business establishments has spread rapidly. In addition to air temperature monitoring, it is necessary to understand these indicators.

6. Understanding energy consumption

During the day, the temperature rise due to solar radiation is significant, and depending on the structure of the building, the temperature continues to rise after sunset.

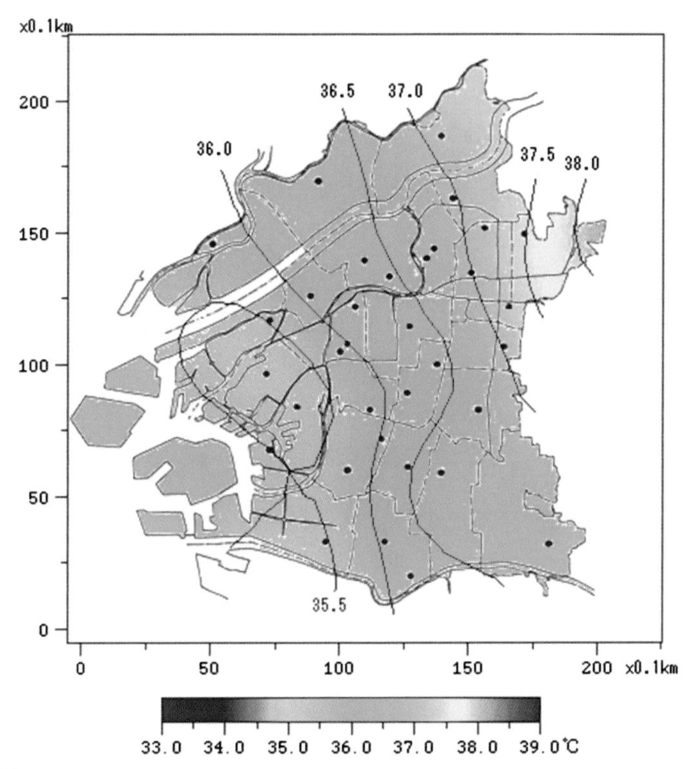

FIGURE 6.1

Contour map in Osaka City of air temperature, August 16, 2007 at 14:00 by 43 monitoring stations. In the daytime, the sea breeze blew from the west to the east and it was cool around the harbor.

However, artificial exhaust heat associated with urban activities, i.e., energy consumption, cannot be ignored. The substantial energy consumption, especially in nighttime commercial and recreational areas, is increasing tropical nights in the center of poor windy cities; reducing exhaust heat is one of the measures to counter this.

In so-called multipurpose buildings that house many small restaurants and bars in the downtown area, there are cases where the outdoor units of air conditioners owned by each establishment are concentrated on the rooftop or veranda. The amount of sensible heat generated at night is large, and the cooling efficiency also is reduced due to high local temperatures (Fig. 6.3). Even if individual store managers are not able to pay attention to the outside air temperature, they can deepen their awareness of the need for heat island countermeasures by being presented with measures to improve the air conditioning system for the entire building that will lead to cost reductions.

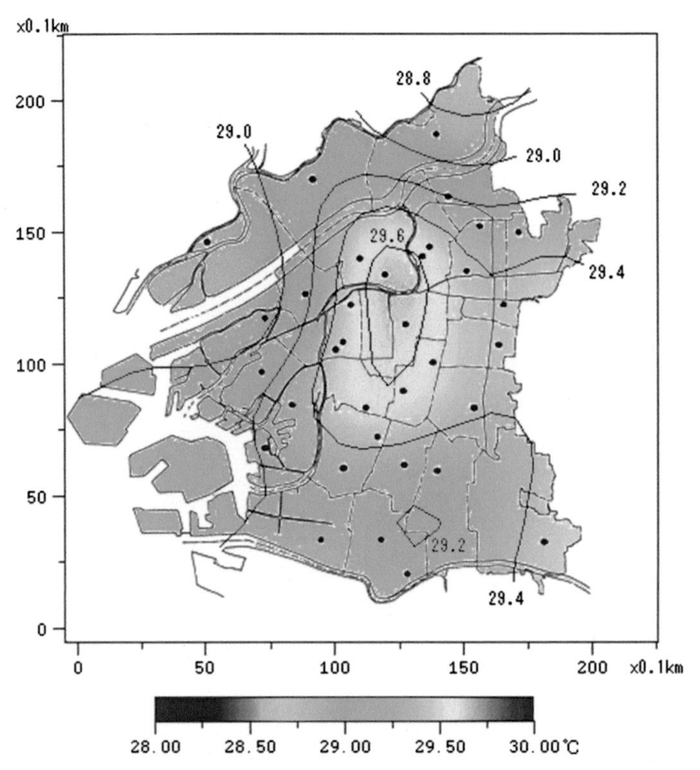

FIGURE 6.2

Contour map in Osaka City of air temperature, August 17, 2007 at 2:00 by 43 monitoring stations. In the nighttime, the wind was calm and it was hot in the central area.

7. Formulating a countermeasure plan in the local government

The "mitigation measures" that suppress the temperature rise in the entire region are large in scale and take time to be effective, so even if the government takes measures, the results are not clearly shown. On the other hand, "adaptation measures" are shown to promptly reduce the health risks posed by local heat pollution, and it is easy for local residents to understand. Therefore, local governments are actively promoting the dissemination of these measures.

"Countermeasure plans" formulated by local governments are generally for a period of 10 years and are reviewed and revised in the fifth year. Reflecting changes in the global situation and new policies by the country, it may be brought forward. In particular, both the Global Warming Countermeasures Plan and the Heat Island Countermeasures Plan cause the city to become hot, so the department in charge of the administrative organization may be the same or the plans may be handled

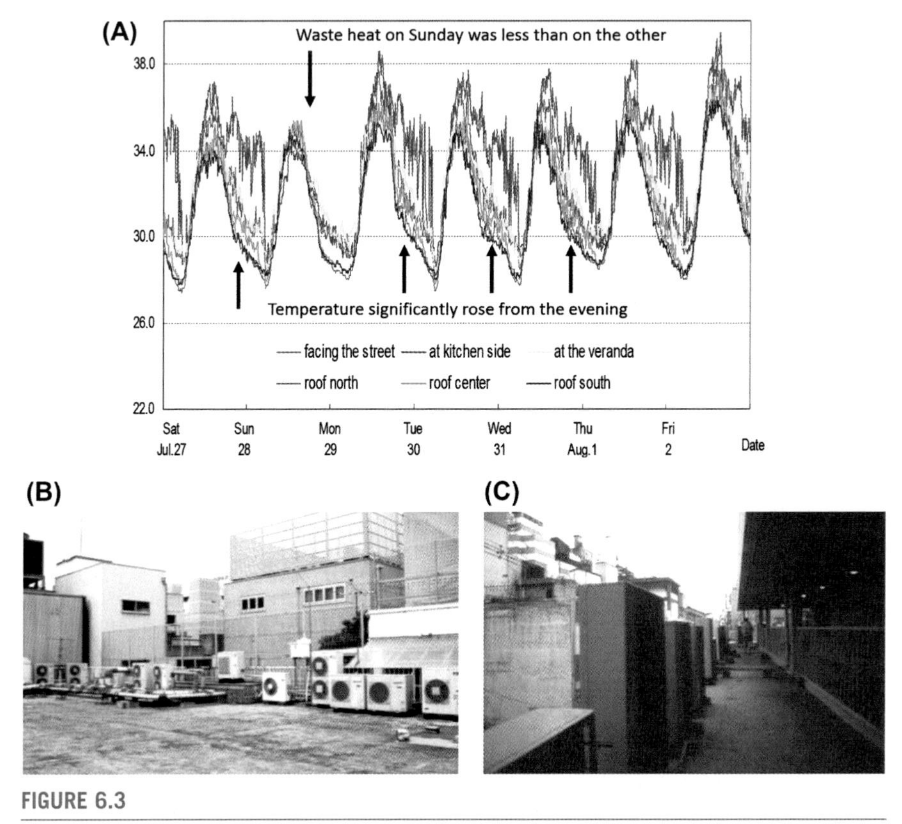

FIGURE 6.3

Variation of atmospheric temperature at the shopping district. (A) Measurement results. (B) Rooftop of the building. (C) Veranda of the building.

together. In particular, local governments with weak financial bases are required to streamline their organizations, and this tendency is strong.

The countermeasure plan includes many demonstration experiments of countermeasure technologies, limited-time model projects, and public awareness events, and the number of projects that provide sustained incentives tends to decrease.

Examples of measures to counter heat island in Osaka City (H29 Business Report): Guidance such as rooftop greening of private buildings; greening facility maintenance plan certification system; promotion of rooftop greening of public buildings; promotion of school greening; operation of bonus system that can ease floor area ratio, such as rooftop greening; guidance for creation of comfortable and eco-friendly buildings; implementation of water retention pavement; energy saving in city buildings; heat island countermeasure business by mist spraying of tap water; heat island monitoring survey; survey research and implementation of

"Wind Channel Vision"; putting up green curtains and carpets in public facilities; development and support for "Osaka Watering Event"; participation in the Osaka Heat Island Countermeasure Technology Consortium (Osaka HITEC); and survey and research on heat island phenomenon and countermeasures.

It is not possible to simply determine the most appropriate countermeasure when it is necessary to evaluate it before implementing the countermeasure, such as which method is the most effective and by how many degrees can it lower the temperature. As mentioned above, it is necessary to consider whether the space-time characteristics of air temperature are different in narrow areas, and whether the necessary requirements according to the actual situation, such as building use and scale, are in place. For example, if you are in the inland, you should cool the living space that has become hot during the day, and if you are in the downtown of city center, you should reduce energy consumption at night. It is important to select additional measures, for example, taking into account the arrangement of windows that can easily capture wind from the coast.

Also, when considering the design of buildings, roads, and street trees of the blocks in the city, not only the application of hardware countermeasure technologies, such as the installation of cool solar shields, in places where the temperature rises due to solar radiation, but also a need for software ingenuity, such as guiding people to a route that allows them to travel while avoiding solar radiation, and guiding people to public facilities where cool sharing is possible.

8. Role of the local government

The construction of large-scale buildings developed by private companies and the redevelopment of city blocks are not designed directly by the government, but there is a system to check whether the contents of construction and development plans for the private sector are appropriate and whether they will result in poor environmental modification. The government conducts assessments that reflect the opinions of academic experts with reference to existing regulations and rules and determines whether they are appropriate. Regarding the countermeasures against heat islands, the following measures can be taken by the government: guidance during large-scale redevelopment (medium to long term); subsidy for introducing countermeasure technologies (time-limited measures); increasing asset value through assessments such as "Comprehensive Assessment System for Built Environment Efficiency (CASBEE)"; dissemination and awareness of countermeasure technologies to increase introduction cases; development (so-called legend) triggered by major events such as the Olympics and World Expo; back-up for a "technology provider" participating in the consortium; and alerting and guiding citizens and business establishments to reduce the risk of heat stroke.

9. Conclusion

Local government tax revenues are declining, making it difficult to continue providing subsidies and incentives to private companies and individuals. Under such circumstances, to promote the introduction of countermeasure technologies, it is necessary to show how the target buildings and blocks improve comfort for owners and residents, and the economic efficiency of energy saving. For this purpose, local governments conducted demonstration experiments and model projects to raise awareness.

On the other hand, heat stroke requires urgent intervention, and the government needs to take the initiative regardless of its financial situation. Especially for the homeless and those surviving in poor living environments, improving the hot environment should be included in the issues of welfare administration. The government is said to be a "vertical organization," but heat island countermeasures must be linked to energy and global warming countermeasures as well as departments that are not in the environmental field, such as construction and urban development. In addition, opportunities for dissemination and enlightenment to citizens are secured through cooperation with ward offices, school boards, and health centers that cooperate with local community association organizations. Also, within the government, the business plan created by each department must be linked to the overall measures. In addition, if a cross-sectional liaison organization is created instead of "vertical division," coordination between departments and securing of budgets can be promoted smoothly.

In addition to organizational collaboration within the administration, the administration must also play an important role in supporting a consortium of private companies and academic experts and working on information exchange and awareness of measures. The Sustainable Development Goals (SDGs) is a global objective from 2016 to 2030, described in the 2030 Agenda for Sustainable Development, adopted at the United Nations Summit in September 2015—a comprehensive approach to addressing a wide range of issues around the world. Moreover, local governments are also developing strategies and promoting various measures related to their goals. Mitigation and adaptation measures against climate change have been promoted mainly by the government; however, as a measure to "ensure the safety and security of life," various private sectors are conducting dissemination and enlightenment in order to deepen citizens' understanding in connection with the SDGs. It is expected that administrative organizations, citizens, and private companies will work together to promote countermeasures.

CHAPTER

Summary

7

Hideki Takebayashi, Dr. [1], **Masakazu Moriyama, Dr.** [2]

[1]*Associate Professor, Department of Architecture, Kobe University, Kobe, Japan;* [2]*Professor Emeritus, Department of Architecture, Kobe University, Kobe, Japan*

Chapter outline

1. Summary of this book

In this book, strategies for introducing adaptation measures are stated as follows.

Chapter 1 describes the background and purpose of this book, with reference to the several precedents. In particular, adaptation city strategy in Karlsruhe was reviewed. Various adaptation menus are organized, and then present and future adaptation cities are specifically presented in the reviewed report. And, efforts in Osaka were overviewed. Initially, the basic directions of measures in Osaka were the following three; (1) reduction of anthropogenic heat release, (2) control of surface temperature rise, and (3) utilization of the cooling effect of wind, green and water. Then, the following was added to them; (4) promotion of adaptation measures. Osaka HITEC (Heat Island Countermeasure Technology Consortium) was established in January 2006, for the purpose of the development and spread of heat island countermeasure technologies, implementation of measures and verification of their effects, and the collaboration between industry, academia, government, and the private sectors. Osaka HITEC started the certification system of heat island measures technology in October 2011. In 2019, the category of certification technology increased to nine; high-solar-reflectance paint for roof, high-solar-reflectance pavement (excluding for roadways), high-solar-reflectance waterproof sheet (membrane), high-solar-reflectance roofing materials (tile, slate, metal, etc.), water-retaining pavement block, external insulation specification

Adaptation Measures for Urban Heat Islands. https://doi.org/10.1016/B978-0-12-817624-5.00007-5

for roof, external insulation specification for outer wall, retroreflective high-solar-reflectance outer wall material, and retroreflective high-solar-reflectance window film. Osaka HITEC is currently working on evaluation and implementation of adaptation measures against extreme heat.

Chapter 2 introduced examples of adaptation measures and their evaluation results. Adaptation measures, such as awnings, louvers, directional reflective materials, mist sprays, and evaporative materials, have been developed with the expectation that they will serve as effective solutions to outdoor human thermal environments that are under the influence of urban heat island. Through several examples of the effects of adaptation measures obtained by demonstrative experiments, it can be seen that shielding of solar radiation to pedestrians is a more effective method of lowering mean radiant temperature (MRT) and new standard effective temperature (SET*), physiologically equivalent temperature (PET). The influences of solar transmittance, solar absorptance, and evaporation rate of adaptation measures on MRT are evaluated by the simple heat budget model. If a shielding device that reflects a large amount of solar radiation and facilitates high levels of evaporation is developed, MRT and SET*, PET will both decrease. And, the evaluation results of the influence on human's physiological and psychological thermal environment were explained based on measured data. Then, examples of adaptation measures at construction site were also described with reference to specific data.

Chapter 3 examined where adaptation measures should be introduced. High-priority areas where is a high need for adaptation measures "hot spots" are explained based on the analysis results of the relationship between urban morphology and radiation and wind environment in the street canyon. For the background of the above discussion, the distribution of air temperature, wind velocity, and thermal sensation indicators of human body are explained on the urban scale. The distribution of hot spot based on solar radiation is dominated by the shadows of the buildings on both sides of the north–south road and southern side of the east–west road. A high weak wind risk area is defined by the road width and the building height in each road parallel or perpendicular to the main wind direction. From examples of hot spot selection, many hot spots are confirmed especially on the north side of the east–west road where the road width is narrow. The number of hot spots in detached house and tenement house district is larger than that in commerce and business district, due to the amount of open space related to weak wind risk.

Chapter 4 examined the specific image of the adaptation city through several case studies. The current state of cool spots and cool roads in Osaka were explained in relation to the factors that cause coolness such as solar shading and evaporation. And, specific adaptation measures, for example, of the use of water, greening, and wind were explained using measured data. Then, the directionality of the urban block redevelopment considering the greening, evaporation, and ventilation was presented with concrete ideas. Ideas competition has been held several times in Osaka HITEC, and valuable potentials are accumulated for future urban block redevelopment studies.

Chapter 5 explained the evaluation method of the adaptation city through some case studies. For evaluation on effectiveness of adaptation techniques to the heat island phenomena, points of view for effects of these on the thermal comfort for pedestrians are required. The thermoregulation mechanisms in the human body and the human thermoregulation model are described based on the relationship between the elements concerning the thermal environment and the responses by the pedestrians. Then, wet-bulb globe temperature (WBGT), new standard effective temperature (SET*), etc. are introduced as the evaluation scales of the adaptation city. Finally, the evaluation methods based on field measurement and numerical analysis are explained through some examples.

Chapter 6 explained the role of local government for realizing the adaptation city. The administrative procedures explained the recognition of actual conditions and risks, the development of targets and countermeasures, the prediction of effects and the implementation and management of countermeasures, and the verification and dissemination of effects. The heat stroke transportation situation, detailed air temperature observation, heat stroke risk awareness, and the relationship with energy consumption are explained as the efforts in Osaka City. Finally, examples of measures in Osaka City are introduced and the role of local governments is described.

Finally, the future direction for the implementation of the adaptation city is discussed as follows.

2. Future direction for the implementation of the adaptation city

From the standpoint that it is necessary to take appropriate adaptation measures in the field of architecture and cities against climate change such as temperature rise expected in the near future, Osaka HITEC has established the "Study Group for Heat Island Adaptation Measures Introduction" as a cross-section group and has been discussing specific images of adaptation cities in addition to the mitigation effects so far. The points of the specific discussion are as follows, and the details are described in the previous chapters; organization of adaptation menu, hot spot analysis, evaluation method of adaptation measures, examination of specific image of adaptation city, role of local government, and cooperation with regional adaptation consortium. Based on these discussions, the practical flow of introducing adaptation measures into the urban block redevelopment in the future is organized as follows.

(1) Examination of adaptation measures introduction place
Based on the current solar radiation and wind environment, local governments, researchers, consultants, and designers discuss about it.

(2) Examination of appropriate countermeasure technology
Local governments, researchers, and manufacturers perform objective performance evaluations of each countermeasure technology.

(3) Predicting the effects of introducing adaptation measures
Researchers, consultants, and local governments predict the effects by demonstration tests and simulations.

(4) Implementation of adaptation measures

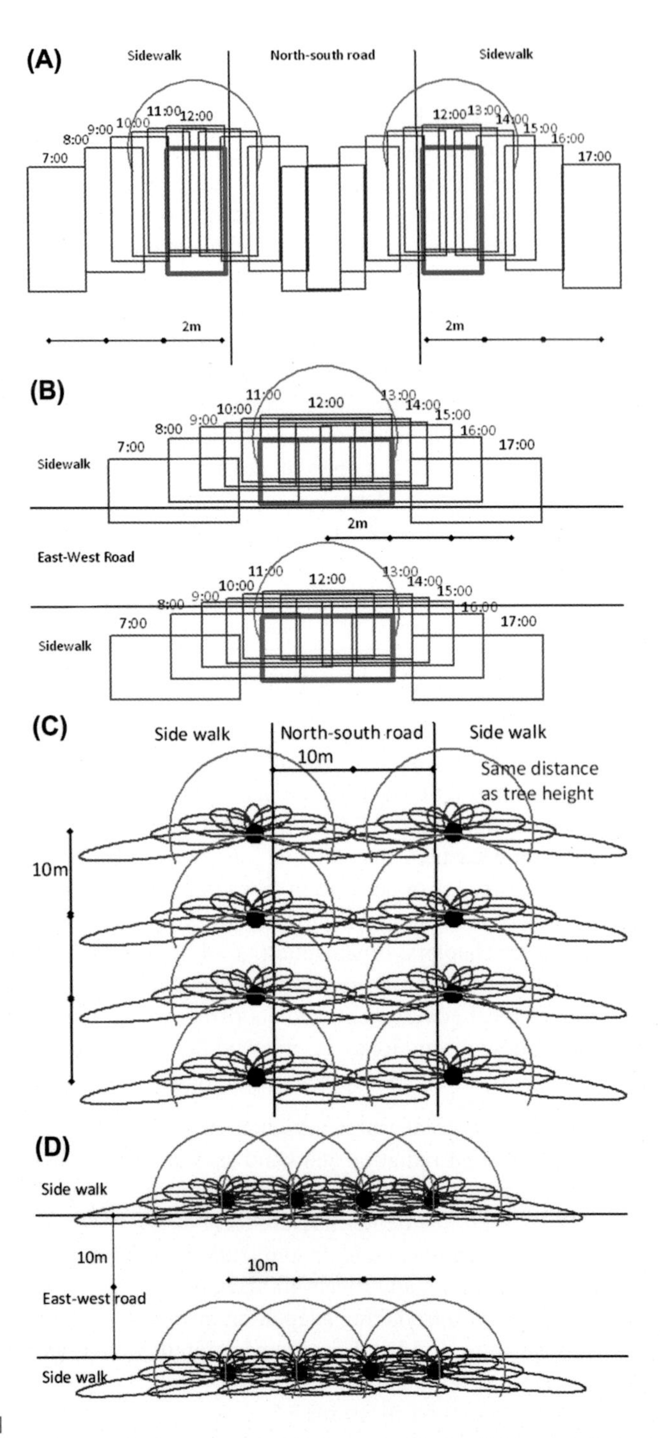
(A)
Sidewalk
North-south road
Sidewalk
7:00
8:00
9:00
10:00
11:00
12:00
12:00
13:00
14:00
15:00
16:00
17:00
2m
2m
(B)
7:00
8:00
9:00
10:00
11:00
12:00
13:00
14:00
15:00
16:00
17:00
Sidewalk
2m
East-West Road
Sidewalk
(C)
Side walk
North-south road
Side walk
10m
Same distance as tree height
10m
(D)
Side walk
10m
10m
East-west road
Side walk

FIGURE 7.1

Companies, consultants, local governments, and researchers implement them into the urban block redevelopment, through social implementation support projects, etc.

The content of this book contributes to these studies and implementations. Lastly, we introduce a practical case study of the efforts of the extreme heat measures in Kobe City.

3. Practical case study of the efforts of the extreme heat measures in Kobe City

Based on the experience of the extreme heat (heat wave) of the summer 2018, Kobe City has been studying and implementing specific measures to target extreme high temperature. In response to a request from the Kobe City local government, the authors implemented the following efforts mainly on the basis of the knowledge of the art.

3.1 Exchange of information and opinions with persons in charge of the local government

We reported from the following viewpoints based on the conventional knowledge.

- Priority introduction places for adaptation measures for extreme high temperature
- Effects of adaptation measures for extreme high temperature
- Hot spots distribution in Kobe City
- Proposals for Kobe City policy

Contents of discussion with persons in charge of Kobe City, the effect of cool spots, and sun shielding by sunshade and tree are shown in Fig. 7.1. (A) In the case of sunshade on the north–south road sidewalk, the sunshade is effective on the western sidewalk in the morning and on the eastern sidewalk in the afternoon. However, the shadow directly under the sunshade is limited to before and after noon. Vertical louvers, blinds, etc. are required to ensure the effect on the directly under the sunshade. (B) In the case of sunshade on the east–west road sidewalk, the sunshade is effective on the northern sidewalk, but it is not effective on the southern sidewalk. (C) In the case of roadside tree on the north–south road sidewalk, the sunshade is effective on a relatively wide area, but it is limited in morning or evening on either sidewalk. The shadow by a tree is unlikely to extend to other trees. (D) In the case of roadside tree on the east–west road sidewalk, the sunshade is effective

Sun shielding by sunshade and tree. (A) Effect by sunshade on the north–south road sidewalk; (B) effect by sunshade on the east–west road sidewalk; (C) effect by roadside tree on the north–south road sidewalk; (D) effect by roadside tree on the east–west road sidewalk.

along the sidewalk. Countermeasure space for extreme heat is formed on the south side of the north sidewalk. The shadow by a tree is likely to extend to other trees in the morning and evening. From the above, it is necessary to arrange trees in consideration of the distance between the east and west directions in order to have a solar radiation shielding effect over a wide area and to obtain many evapotranspiration effects. Conversely, a countermeasure space for extreme heat is formed when trees are arranged side by side in the east–west direction.

3.2 Providing information and exchanging opinions to administrative personnel in each department of Kobe City

An expert collaborative workshop was held by administrative personnel in each department of Kobe City and experts in urban environmental engineering of Kobe University and other universities. The workshop program is as follows.

- Greetings by Director of Urban Strategic Research Office, Planning and Coordination Bureau of Kobe City and Vice President for Industry-Academia Collaboration of Kobe University
- Special Lecture by Professor Emeritus Jürgen Baumüller of University of Stuttgart, Germany, with the following title: Implementation of adaptation measures for extreme high temperature in German
- Report by Prof. Hideki Takebayashi of Kobe University and Mr. Ushio Tozawa of Kobe City, with the following title: Investigation for extreme high temperature countermeasures in Kobe City
- Discussion for implementation of measures to adapt to extreme high temperature and Summary

As an example of the study results, less and more than 80% of the total daily solar radiation on a clear day in summer is shown in Fig. 7.2. After reporting the results of the study on countermeasures against extreme high temperature in Kobe City, administrative personnel in each department introduced the related efforts and confirmed the possibility of specific measures. It was pointed out that the redevelopment plan in front of Sannomiya Station, which is the main station in Kobe, should reflect the knowledge about measures against extreme heat. The effectiveness of collaborative discussion and examination between the university and the local government and participation from related departments was recognized.

3.3 Cool spot demonstration experiment

A demonstration experiment of cool spots on the outdoor space carried out from July 3 to September 30, 2019, where fractal sunshades with fine mists were installed on the plaza in front of the famous department store and the North-South street in front of Sannomiya station. An example of a cool spot set up by Kobe City is shown in Fig. 7.3. The mist worked from 11:00 to 15:00 on Fridays, Saturdays, Sundays,

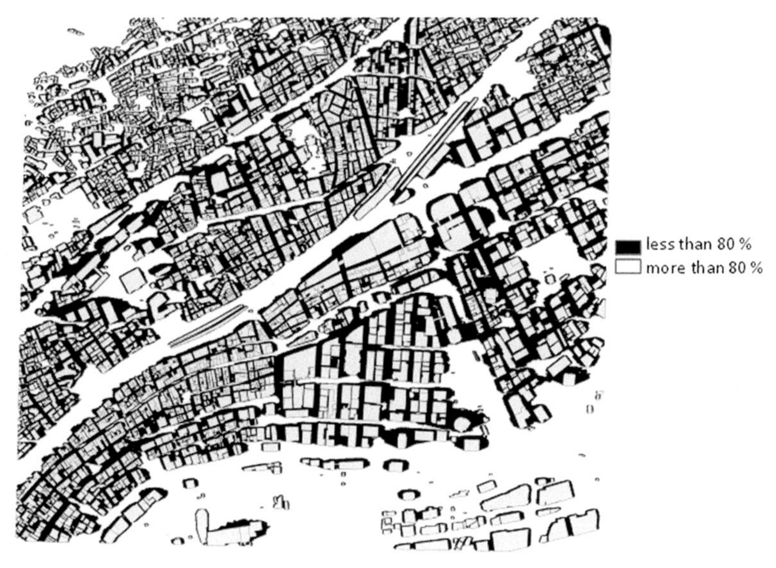

FIGURE 7.2

Less and more than 80% of the total daily solar radiation on a clear day in summer.

FIGURE 7.3

An example of a cool spot set up by Kobe City.

and public holidays. The effects of creating rest and bustle were also investigated as well as the effects of improving the thermal environment.

3.4 Symposium on countermeasures against extreme high temperature

A symposium on countermeasures against extreme high temperature "Wisdom and ingenuity to survive summer" was held at Kobe headquarter building of Sumitomo Mitsui Banking Corporation on July 5, 2019. There were about 200 participants, including citizens and government officials. The program is as follows.

- Keynote lecture by a hospital director, with the following title: Measures against "heatstroke."
- Research reports by several professors, with the following title: Feel "wind" to prevent heat stroke, measures against extreme high temperature focusing on urban forms, measures against extreme high temperature in the indoor space, clothes effective for heat stroke prevention.

In the panel discussion, the coordinator, Mayor of Kobe City requested each panelist to provide ideas and information on specific measures against extreme high temperature, and the panelists presented ideas and knowledge.

We must continue theoretical examination and practice in society.

Index

'*Note*: Page numbers followed by "f" indicate figures and "t" indicates tables.'